SCIENCE, REMOTE VIEWING, AND E S P:

BEYOND EINSTEIN'S HORIZON

by

Ken Renshaw

Constellation Press

Cambria, California

Library of Congress Control Number: 2009907568

For information address:

Constellation Press
1790 Ogden Drive. Cambria, CA 93428
krenshaw@remoteviewesp.com
ISBN: 978-0-9794000-3-2
Published in the United States of America

To

Joycee,

my muse

Acknowledgements

Special thanks to Russell Targ and Elizabeth Rauscher for their encouragement and suggestions.

Jacqueline Parker did the heroic job of checking all of the references at The Library of Congress and formatting them into the MLA standard.

Thanks to Gayle Oksen for her generous review and comments.

Table of Contents

If you use your eyes, you can only see as far as your three-dimensional horizon.

If you use your power of four-dimensional visualization, you can only see as far a Einstein's horizon.

However, our reality is eight-dimensional.

If you use your consciousness, and look in the right eight-dimensional direction, you can see anywhere in space or time, past or future.

SCIENCE, REMOTE VIEWING, AND ESP:

BEYOND EINSTEIN'S HORIZON:

Einstein overturned 200 years of physics with his Theory of Relativity. In doing so, he established a 4-dimensional horizon view of the world that has constrained popular scientific thinking. It is possible to explain remote viewing and ESP by going beyond his horizon into an 8-dimensional complex Minkowsky space.

I Introduction

George was leaving my office after discussing a new, innovative antenna for the communication satellite we were designing. George was one of the technical superstars of out organization. His specialty was electromagnetics. George had a couple of PhDs in the area and was enjoyable to collaborate with. I said to George, "Sometimes I feel vibrations on people that seems to be somewhat related to their emotional state. Do you have any kind of idea what it might be?" George looked incredulous, and I knew that I had just said something that was outside his scientific belief system. I got the same reaction I would have if I said that I had been talking to aliens and riding their space ship. George said, "There is nothing in electromagnetic theory that would explain that." I observed that his vibration had changed and dropped. I changed the subject to the antenna we had been discussing. After a while George left and I thought he was a little upset with me for talking such nonsense. On my business card it said

Chief Scientist: I was supposed to be scientifically conventional.

Two hours later, I was in another meeting with a group of specialists. I was leading a proposal effort for a communication satellite we were proposing to a telephone company. We would spend about a million dollars in two months to write a proposal about a foot thick. It would have contributions from a couple of dozen departments dealing with some subsystem of the satellite or its construction, test and launch. Two of the three men from the department had PhDs related to this fairly narrow specialty. In our meeting we were to decide on topics this department would contribute to the proposal. They didn't have much of an idea what would be important to describe or how much effort should be expended. They were specialists. They had a list of about a dozen topics that they thought could be included. Five of them were obvious to be included. We went through the rest of them, cutting the topics that might not be of interest to the customer. Finally, we had narrowed the list to seven topics. As we went through each topic, I could feel the team leader's vibration change. I suggested that we could delete the seventh and the leader's vibration abruptly dropped. I knew we had cut far enough and then put the item back on the to-do list. They went away happy with their assignment.

Later, I talked to another customer, Warren, in Australia. Then, there were telephone circuits via communication satellites over the Pacific Ocean. The signal went from California 19,000 miles up to a satellite in space and then down 19,000 miles to Sydney. Even at the speed of light, there was about a quarter of a second between the time I said something and Warren heard me. Then, there would be a quarter second after Warren replied before I heard it. I knew Warren quite well and could "tune into" his normally high vibration. Warren said, "have you got any idea how much that study we talked about might cost?" I replied, "About a million dollars," A quarter second later, I felt Warren's vibration drop to an angry level and then a quarter second later I heard him say, "Bloody hell! I don't have that kind of money." Somehow, the vibrations traveled faster than the speed of light.

This kind of phenomenon violates the known laws of Physics. It is truly "Outlaw." According to Einstein, nothing can travel faster than the speed of light. It also violates Maxwell's Equations (which we will describe in a subsequent section) that govern propagation of radio waves or light. It is impossible for a human body in California to radiate enough energy to send any signal all the way to Australia. Furthermore, physiologists can not identify any sensory system in the human body capable of receiving or transmitting vibrations. This is truly an "Outlaw" experience, to people with George's scientific belief systems.

George is a theoretical scientist that deals mostly with equations. He uses equations to give theoretical predictions of performance of antennas and things. George believes there are fundamental laws of physics. Anything that does not fit those laws of physics is a crackpot idea.

Any phenomena not explainable by the widely held scientific paradigm are rejected as superstition, no matter how convincing the scientific data presented. One group of credible researchers at a credible research establishment does some sort of ESP experiments according to impeccable scientific protocols and publishes the results. Another group of scientists vehemently contest the results.

While working in a scientific environment, I learned that there were two topics that are dangerous to discuss with engineers or scientists. The first topic is religion. The second is any sort discussion of ESP or any other sort of psychic phenomena. Somehow, many engineers or scientists have strong beliefs for or against these subjects.

Any discussion of these topics would lead to what I call the "Bigot's Protocol." If the topic is psychic phenomena it goes like this:

1. A person describes some sort of psychic phenomena or experience to a Bigot.

2. The Bigot displays anger, stops listening, and reasoning. and becomes threatened.

3. The Bigot becomes righteously indignant.

4. The Bigot explains that what you described is irrational and incompetent. They invalidate you as a person and say nobody with any sense should think like that. They try to convince you that you are an idiot.

5. The Bigot will try to destroy your professional or social status.

6. The Bigot will be intolerant of your attempts to communicate with them.

It has surprised me how many otherwise competent, friendly people will run the Bigot's Protocol when the subject of any kind of psychic phenomena comes up. Somehow, the subject touches a deep seated core belief.

When I was at the University, I was very much the same way. As a student, I believed what I was being taught. I also felt strong emotions and vibrations but always assumed they were mine. In later life, I realized that some of these vibrations were from other people. I had to accept the idea that, although there is no scientific rational for sensing other people's vibrations, I could do it. Somehow, I had this ability.

I am not alone is feeling unexplainable vibrations of people. My wife and some of my children feel people's vibrations. I have many friends who also feel vibrations. I believe elementary ESP is a native ability of most people that can be trained and developed.

Vibrational perception is not limited to simple emotional level vibrations. The excitement at a rock concert or a sports event is a related vibration. Actors or other performers "feel" their audiences, particularly during applause.

Great musicians in great orchestras "feel" the emotional vibrations that the conductor and the composer wished to express. Audiences similarly "feel" great performances.

There are too many aspects of Extra Sensory Phenomena to go into here. I did a Google search on "ESP Extra Sensory" and got 837,000 hits. An Amazon book search of "ESP" and got 70,000 hits. There are obviously millions of people who believe in various forms of ESP. Also, there are millions of people who adamantly believe ESP is nonsense. For those of scientific training, no amount of scientific data from experiments run according to the strictest scientific protocols will change their belief systems on ESP. The subject falls into the category of "superstition" and proper scientific people do not deal with that.

Although my business card said, "Chief Scientist," I have always been suspicious of scientific conclusions deduced from experimental or analytical studies. I lost my belief in the purity of Science while working in one of my first jobs after graduating from the University.

War Games Demonstrate Limitations of Science

Most people think the scientific method is a pure methodology that identifies and validates truth. In reality, the scientific method creates only consistent sets of data to support a belief that some may hold. Frequently, highly qualified scientific expert witnesses in trials testify for both sides of an argument. One will present scientific data that shows that a drug kills people. An opposing expert witnesses will claim that there is no valid scientific data showing that the drug causes people's deaths. Both sides may have peer-reviewed papers and extensive data to support their beliefs.

My understanding of the scientific method was formed in the late 1960's when I worked as a consultant to the Navy in Washington D.C. The belief that weapons systems should be "cost effective" was in-vogue in the Department of Defense. The Navy believed it should have lots of ships for Naval officers to take turns commanding. When the Navy wanted to buy more ships, (especially aircraft carriers that are really fun to drive) it would have to do scientific studies to show that buying more ships was more cost-effective than doing something dumb like buying Air Force fighters. An Admiral, who knew in his heart that the Navy needed more ships, would call in a study contractor, one of the so-called "beltway bandits" who consult in Washington, and say, "Do me a study that justifies buying another aircraft carrier." Nine months and several million dollars later the contractor would deliver an eight-inch thick study report that was based on war game simulations. The study would conclude that a new carrier would be more cost-effective than a bunch of Air Force fighters or bombers.

Most of the people I knew who performed and used these studies really believed in them. They proved the hypothesis that buying nuclear carriers was more cost-effective than buying airplanes.

I worked for an Admiral whose job it was to make sure that the studies had integrity, "held water" to use a Navy term. I would read the studies and find out which key assumptions or methodologies in the war games were necessary to have the answer turn out in the Navy's best interest. I would identify these so that the Navy would not go to the Department of Defense with a study justification that was easily disputed.

I told a statistician who worked for me that this whole process was a waste of taxpayer money, prostitution. He corrected my thinking. The method was Newtonian. What the Navy did illustrated the scientific method. First, a hypothesis was formed based on intuition and general knowledge. (The belief that another nuclear powered carrier was needed.) Then, a methodology and data were selected (by the beltway bandits) and the hypothesis was proven and published. Then, the study was peer-reviewed by me (I was also a beltway bandit) and the Department of Defense and their (beltway bandit) consultants. If the hypothesis were proven and the belief accepted, the Navy got a few billion dollars for a new carrier. It was all very scientific.

Unfortunately, there were high level scientists and managers in the Department of Defense who treated the war in Vietnam as a set of theoretical war game problems. They were trying to optimize sets of equations that they thought described the war. A syllogism might be, "If we conduct 'Operation X' and we have a thousand casualties but causes the enemy to have fifty thousand casualties, we are winning. 'Operation X' is therefore, a good idea and if we do it as an experiment we can determine the 'exchange ratio' coefficient that we can use in the overall war simulation."

My daily taxi ride from my downtown Washington office to the Pentagon took me by the Arlington cemetery. I observed the cemetery filling up rapidly. One day, I observed a stooped-over, crying, grieving family walking away from their son's burial. I thought, "Those people in the Department of Defense are madmen." I immediately started changing jobs and working my way to the commercial arena. I never worked on a weapons system project where the product killed people again.

I left Washington with a different view of Science than most people. Science is a belief system, and not a set of absolute truths.

Science Is A Belief System

Among those who are not involved in science, their basic education in grades 1-12 has conditioned them to be suspect of anything that should be explainable by science and isn't.

Few people think of science as a belief system in the same sense as a religion. Scientists tend to think they are dealing with truth rather than beliefs.

In the dark ages' first millennium, in the Western countries, everything was a Mystery. When something happened, such as your cow dying, you probably wouldn't be able to assign a cause. Superstitions and rituals were supposed to act as causes to prevent certain things from happening or cause other things to happen. All in all they were pretty unreliable.

Then, at the beginning of the second millennium, organized Church emerged and said, "We will take over the management of Mystery." There was then only one cause, God's will. God micromanaged everything.

Then, after the middle of the second millennium, people like Copernicus started a revolution. They said there were physical laws which were causes of things happening. Then, the Age of Enlightenment began in the late eighteenth century. Rational thought or reason, replaced Mystery. Science flourished in explaining causes. Science would explain everything

God became the master planner and not the day-to-day arbitrator of human events.

Today, if there is some unexplained event, we call the scientists first. When a Middle Eastern airline pilot decided to commit suicide by diving an airliner into the ocean, scientists, not clergy, were called to investigate.

Today, society has generally forgotten that Science is a belief system, a collection of beliefs. In many ways, scientific and religious beliefs are obtained in the same way. In our early training, and indeed through grade 12, we are taught Science. We are told everything in science is true, and everything in the "Holy" texts of science is absolute. Texts are approved by state officers and committees who canonize the texts. Outdated texts are discarded as apocrypha. We are taught the scientific techniques by using simple experiments with simple phenomena with easily proven results. We come to believe that everything is amenable to proof by experimentation, and only things provable by theory and experimentation are true.

When I discussed vibrations with George, I was in conflict with his belief system. There was nothing in his "holy" texts on electromagnetism that could explain the phenomena. When I talked of vibrations, I became an infidel.

The people I described in the Department of Defense during the Vietnam war believed in scientific management. They believed, if something is scientifically shown to be more cost-effective than something else, their Congressional testimony should reject any other idea as heretical. These beliefs have continued to the present.

The people doing scientific war game studies really believed in them. Their reports were orthodoxy.

Most people believe that Science is absolute in its truth. No other religion has this undisputed franchise. Somehow, Science as a method of inquiry and as a body of knowledge can have no flaw. Any attempt to convince a card-carrying member of todays "Church of Science" of something that does not fit their paradigm runs into a solid barrier. That barrier is Science's belief that their God (Newtonian-Einsteinian-Quantum Science) is the only true God.

From personal experience, I knew that my ability to sense vibrations from people in distant places violated known laws of physics. I always assumed that the flaws were in Einstein's theories of relativity.

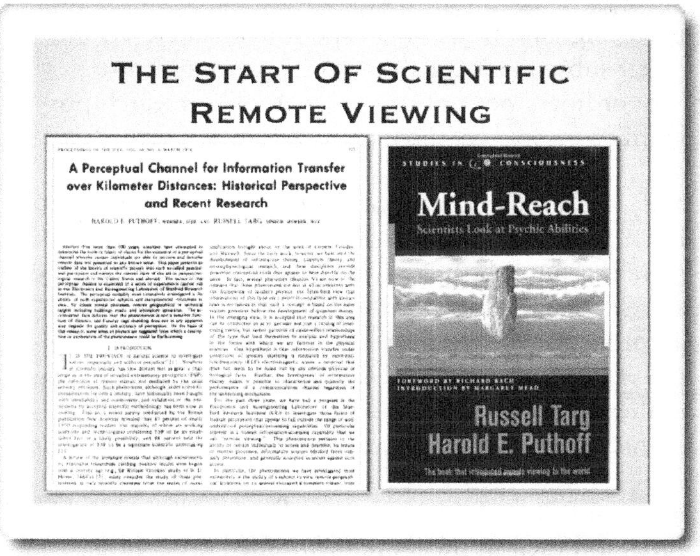

The Stanford Research Institute Program

I learned about the Stanford Research Institute, SRI, Remote Viewing Project in 1976 when I read an article[1] in the peer-reviewed Proceeding of The Institute of Electrical and Electronic Engineering. I was excited about that article because it showed that remote viewing, and ESP phenomena, could be experimentally shown to be valid. Further experimental results appeared in the Proceedings later in that year.[2] A series of letters to the editor appeared after that from skeptics denouncing the research as unscientific and unworthy of publication in the Proceedings. Some resigned their membership in the Institute. I recognized the Bigots' Protocol was being run by non-believers of ESP.

Then, in 1977 The SRI experiments were further revealed in the best-selling book[3] *Mind-Reach: Scientists*

Look at Psychic Abilities. The program of remote viewing started with the New York artist, Ingo Swan, contacting Dr. Hal Puthoff at Stanford Research Institute regarding remote viewing experiments in New York. Their subsequent collaboration in demonstrations of extraordinary perceptions led to a 1972 research program sponsored by the CIA.

SRI, was, affiliated with Stanford University and was not a place where people are allowed to do flaky research or publish the results of questionable quality research. The remote viewing projects were conducted according to the strictest scientific protocol by recognized, credentialed scientists. SRI required independent evaluation of the results of the remote viewing experiments by researchers from other unassociated projects.

The results of experiments were first published in Nature[4] and then in the afore mentioned proceedings, and then in *Mind-Reach*. That book gained the interest[5] of the Intelligence and Security Command of the U.S. Army. They wanted to know the vulnerability of classified Army facilities to spying by remote viewing. They started a parallel but coordinated program at Fort Meade, Maryland.

Both programs evolved into highly classified remote viewing intelligence gathering programs which continued until 1995.

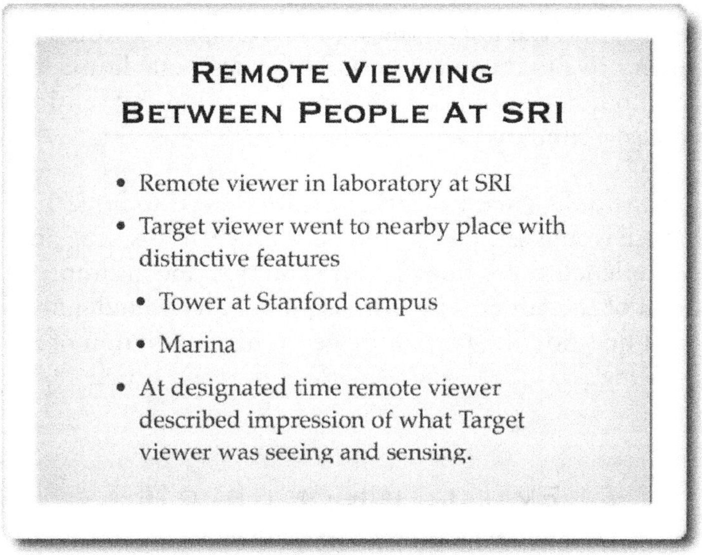

The original SRI experiments were very simple. An experimenter selected a number of target locations within twenty minutes drive of SRI. Target locations were selected to have strong visual or other sensory clues: a tower, a bridge, a busy shopping mall, a boat dock, etc. The experimenter who selected the target sites wrote out driving directions to each location and placed them in separate sealed envelopes. The target direction envelopes were placed in the SRI director of research's secure safe. Since only he had access to the safe, no one could say that the remote viewers had somehow visited the target site before the remote viewing.

Three people would begin the experiment in an ordinary SRI office. One would be the remote viewer, and another would be the "outbounder" and go to a target. The outbounder would go to the director of research's office. One of the sealed letters of driving instructions would be randomly drawn from the director's safe.

The outbounder would go to his car, drive a couple of blocks and then open the sealed envelope. He would then drive to the target area. At the designated time, the outbounder would concentrate on viewing and sensing the surroundings.

At the designated time, the remote viewer in the SRI office would sense what the outbounder was seeing and experiencing and draw picture and describe his impression of the target. The third experimenter would guide and question the remote viewer in his exploration of his sensing.

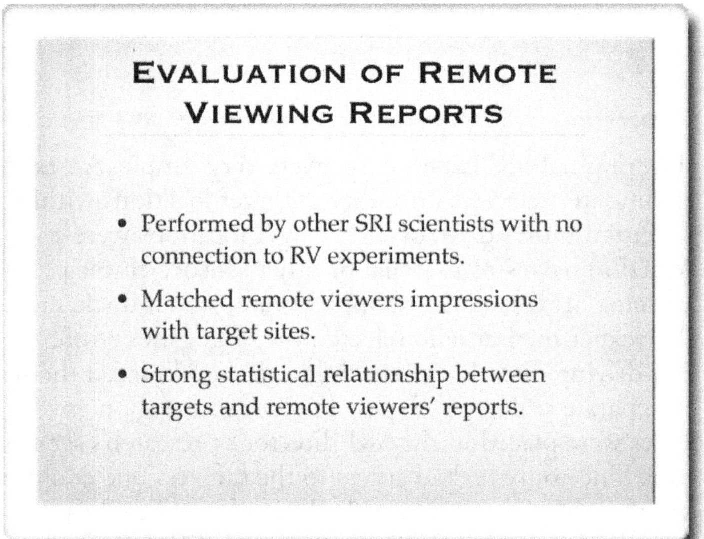

After a significant number of remote sensing trials, a panel of SRI scientists with no connection to the remote viewing project would visit the target sites and judge the reports. At a particular site, they would go through all the remote viewers' reports and identify descriptions which matched that site. A statistically significant number of matches were shown.

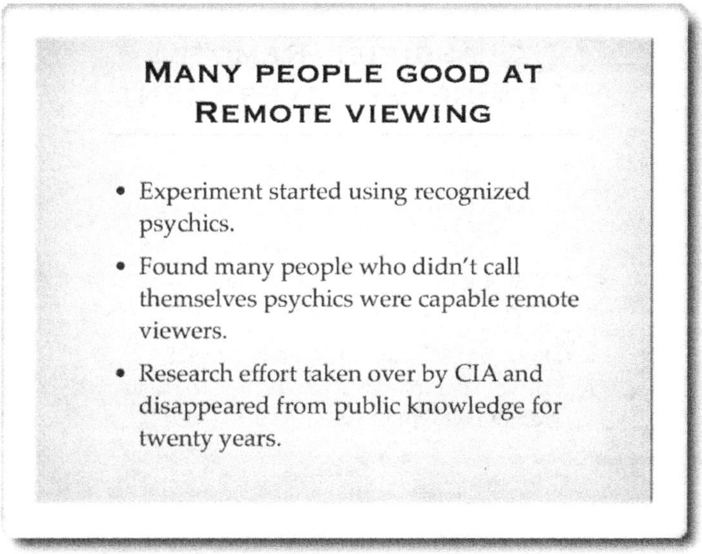

It was found that the ability to remote view sites was fairly common. The experiments started with people with reputations of having psychic abilities. It was then found that many of the SRI staff, including office workers, could do remote viewing. One visitor from the CIA asked for a demonstration. The researchers had the visitor do the remote viewing himself. He did well in describing a target. Soon after that, the CIA funded the project.

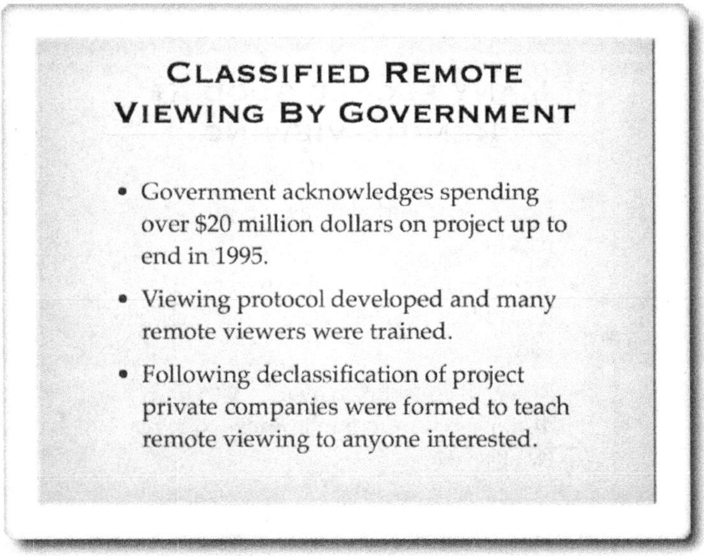

The government acknowledges spending over twenty million dollars on remote viewing programs up until 1995. Protocols for screening and training people with good aptitudes for remote viewing were developed, and procedures for remote viewing of targets and evaluation of the results were formulated.

Examples of SRI Remote Viewing Experiments

Because of the highly classified nature of the program, few examples of the results are available. There are four examples available from Russell Targ's espresearch.com website.

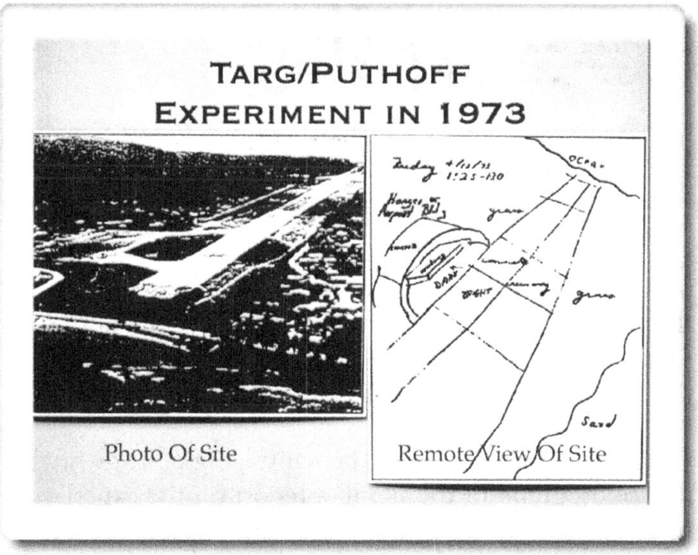

In 1973, Hal Puthoff was traveling through South America. The experimenters at SRI did not know the details of his itinerary. At a designated time each day, Hal would go to some convenient viewpoint and study and photograph the scene. At SRI, the psychic, Pat Price, would remote view the scene and make sketches and write descriptions of his perception. On the day that the above experiment was to take place, Pat suddenly became ill and Russell had to take his place as the viewer. In the above figure, it is seen that Russell's drawing closely agrees with Hal's photograph. Russell even correctly saw the "ocean at the end of the runway."

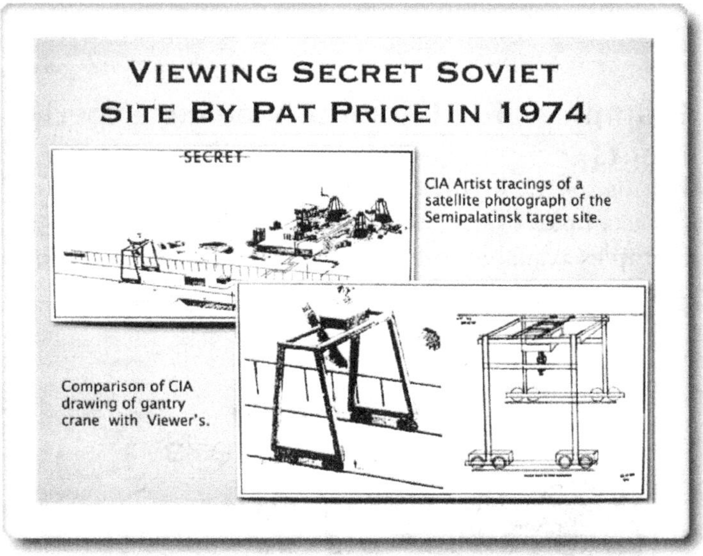

A later 1974 experiment, done for the CIA, used Pat Price to view a Secret Soviet site. Pat drew the rather detailed picture of the crane shown in the lower right. Note that he identified it as having eight wheels.

The CIA later provide the pencil tracing of a spy satellite photograph of the site in a report on the experiment. (They did not show the actual photograph because it might reveal the highly classified resolution of the satellite images.) The detail of the satellite image of the crane is shown next to Pat's drawing. Pat also described the activity going on inside the building underneath the crane. His observations were later verified to be correct.

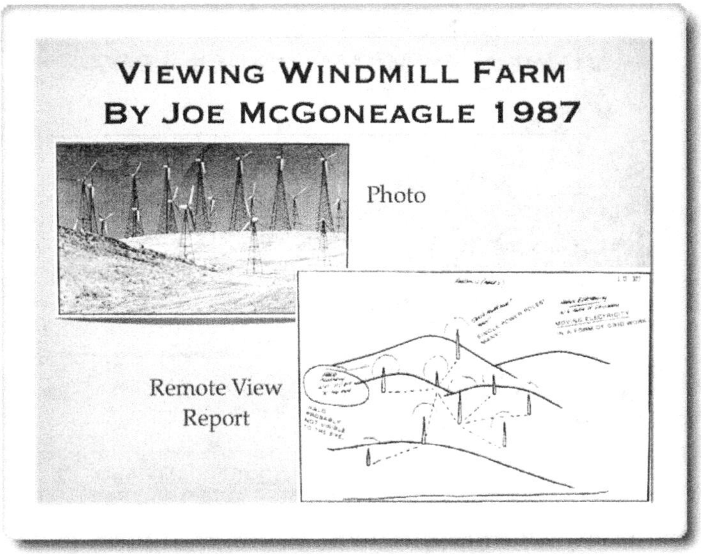

In 1987, Joe McMoneagle, whose ability as a remote viewer is described in Section IV, did experiments with Edwin May at SRI on remote viewing the location of an unknown "agent." The "agent," actually a US Government employee, was unknown to Joe. Joe was given only an index card with the "agent's" social security number written on it. At a designated time, Joe was to view the location of the "agent." In the above experiment, the "agent" was at a windmill electric generating farm about a hundred miles from Joe's location. The photograph of the "agent's" location is compared to Joe's remote viewing sketch. In this experiment and in the one that follows, the photographs were only shown to Joe long after the experiment was run.

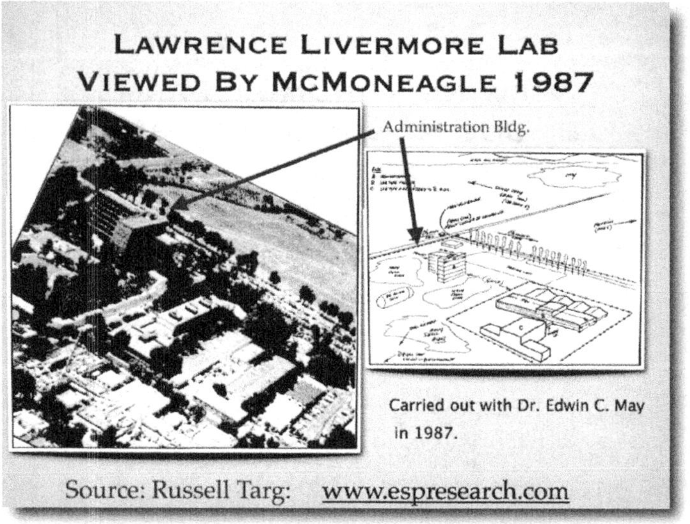

In this experiment, run soon after the windmill experiment, the "agent" was on the top floor of the six story Administration Building at the Lawrence Livermore National Laboratory. Joe had never been to this location. On the right in the above diagram, Joe's sketch of the Administration Building and the surroundings is shown. Joe also correctly perceived the location of the "agent" in the building on the sixth floor.

These four remote viewing examples were not highly classified and have subsequently made their way into the public domain. The remainder of the government twenty years of remote viewing experiments and operational results remain classified.

Revelation Of Classified Remote Viewing Program

When the subject of remote viewing disappeared from technical literature, I assumed that it had somehow disappeared into the classified world. I first learned of the official existence of the intelligence remote viewing program on November 28, 1995 when Ted Koppel broke the story on this ABC TV *Nightline* show[6]. Soon, selected government reports were declassified. Many critics described the program as of little value. Criticism of the program's effectiveness was voiced by many Government officials.

In this kind of scenario, it is difficult to discern what is damage control by the CIA, misinformation to protect the security of ongoing programs, red herring disclosures to protect classified data from Freedom Of Information Act requests, or factual disclosures. The revelations about the government's program must, therefore, be taken with a grain of salt. Intelligence agencies, by their very nature do not always tell the whole truth. The fact that the program lasted over twenty years indicates some useful results. An official CIA sponsored evaluation study was declassified. It said the remote viewing program was of little operational intelligence value. Declassification of parts of the program amid evaluations that it was of no use could have been intended to hide the fact that there were parts that were highly successful. This may be like telling the children's story of "The Three Little Pigs" and leaving out the part about the third pig. Revelations about scientific results from The Intelligence Community cannot be taken at face value.

Principals involved in the program, Hal Puthoff and Russell Targ, then wrote articles[7,8] which appeared in the Journal of Scientific Exploration. Edwin C. May, the project director at SRI International, also wrote a commentary[9] on the evaluation study used to kill the program. He blamed the politics of interagency fighting. He also noted the evaluation study participants did not have access to the results of the whole program nor have the security clearances to access the remote viewers or their operational intelligence product. The study was rigged to produce the forgone conclusion that the program should be killed.

The government created two schools of thought regarding remote viewing. First, there are those who participated in the program and know the facts about how effective the program was. Because of security constraints, they cannot, even today, disclose the details and value of the operational intelligence gained through remote viewing. They can say the program had great results but can provide little data to support the conclusion.

Secondly, by crafting an evaluation study rigged to conclude that remote viewing had little intelligence value, the Government could discredit the program and provide the ammunition for anybody to prove the program was ineffective. I might add that this negative reporting on the effectiveness of remote viewing would also provide a disincentive for any other nation or terrorist group to start their own remote viewing program.

I had always believed that remote viewing was a real, verifiable phenomena, but never thought it would be something I would engage in. I forgot about the subject. But, in 2007, I attended two lectures on remote viewing at the Science and Consciousness Conference In Santa Fe, New Mexico. In the first lecture, I found that I could do well in the simple exercises offered. I also attended a lecture by Russell Targ and bought his book.

Russell Targ's guide to remote viewing[10] gives a summary of the twenty year history of remote viewing. It also presents the case for the scientific validity of remote viewing and related phenomena such a distant medical diagnosis and precognition. In the book he says, "To me, these data suggest that all space-time is available to your consciousness, right where you are."

The AMAZON description of the book reads:

> "....illuminates the phenomena of intuitive medical diagnosis and distant healing in a groundbreaking synthesis of research and empirical data. Drawing on a broad range of spiritual traditions, Targ demonstrates that these psychic abilities offer a path of self-inquiry and self-realization and have the power to expand each person's limited awareness into the consciousness shared by all

> beings. Targ explores the scientific and spiritual implications of remote viewing, as well as offering practical techniques and exercises to nurture this universally available but often untapped skill."

The book has sold over 100,000 copies and is internationally recognized as an exceptional introduction to the study of consciousness and psychic abilities.

Personal Experience In Remote Viewing

My interest in remote viewing returned after I read the book. I started researching all the literature on remote viewing. One paper[11] by Elizabeth Rauscher and Russell Targ dealt with how extrasensory perceptions could appear to travel faster than the speed of light. That really sparked my interest and I proceeded to read many complicated papers in physics and mathematics on the subject.

I found out that Russell Targ was giving a weekend seminar on remote viewing at Esalen in Big Sur, California near where I live. My wife, Joycee, and I signed up for the workshop. Joycee, who had been a kindergarten and art teacher, had never known anything about remote viewing and does not share my scientific interest in the subject. I was surprised that she did quite well in the exercises.

Russell Targ's first exercise for us was to remote view an object that he had in a duffle bag. While he concentrated on making a mental picture of the object, we concentrated on him and what was in the bag. We drew and wrote our impressions of the object. Then, the object was revealed to us, a long handled kitchen tool for

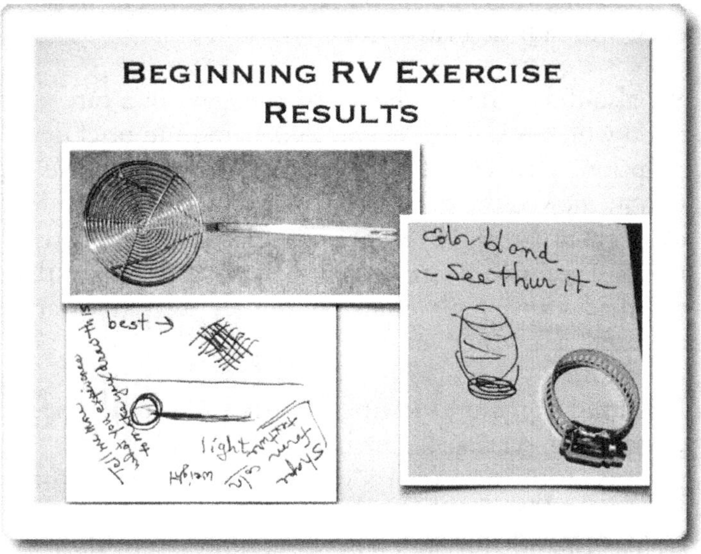

straining something like pasta. In the above figure, Joycee's impression is shown along with a photograph of the object.

We later repeated the exercise with a partner who brought in something in a paper bag they had selected as an interesting object. Joycee's partner had brought a pipe clamp, which is something she was not at all acquainted with. He held a picture of the clamp in his mind and Joycee drew and wrote down her impressions. The clamp and her impressions are also shown in the figure. Her drawing is amazingly close.

I brought a highway flare in my bag. I imagined the flare burning and my waving it to warn traffic. My partner immediately drew pictures of something exploding in a cone and said it was fire. She also said that the fire was where something bad had happened.

Most of the thirty-five members of the class had similar results with remote viewing objects in the bags. It seemed like a very common talent that could be developed with a little instruction.

We also did exercises where a partner gave us a card upon which was written a person's name and brief description. That served as the guide to the person. We remote viewed the person's body and sensed any medical problem areas they might have. I was quite surprised that as I scanned the remote person's body, I sensed that they had a knee replacement and other problems that my exercise partner verified.

I believed in remote viewing for medical intuition because, a few years ago, I had a remote viewing telephone consultation with Mona Lisa Schultz in Boston. She is an MD and PhD, who is a consultation neuropsychiatrist and also does remote viewing as a medical intuitive. She asked me only my name and age. She then moved her attention through a my body, "saw," and identified anomalies that I knew of from seeing local doctors. She then sensed my mental vibrations and identified issues that appeared to be influencing my health. She then recommended eliminating specific activities that were causing stress in my life. She was spot-on in her observations and recommendations. She has published several books[12,13] on intuitive medicine.

After the Esalen workshop, I decided to write this book. I had to figure out how to present a lot of very mathematical ideas without getting really boring. There is also scientific prejudice to be overcome in explaining remote viewing. I think the skepticism is exemplified by the page on remote viewing in Wikipedia[14].

The article illustrates the divergence of belief systems on the legitimacy of remote viewing. Wikipedia articles can be submitted by anyone and be edited by anyone else. In this article, those with opposite belief systems have edited and added to each other's work. The article gives a good history of the research and the opposition to the research. It also indicates current significant opposition to the idea that anything like remote viewing or ESP could have any scientific validity.

The article was an excellent example of the Bigot's Protocol at work. I was reminded of the Scopes Trial of 1925. That was big league bigotry. I decided to set up the explanation of remote viewing in a trial metaphor.

---

[1] Puthoff, H.E., and R. Targ. "A perceptual channel for information transfer over kilometer distances: Historical perspective and recent research.." *Proceedings of the IEEE.* 64.3(March 1976): 329- 54. Print.(Available as a download at espresearch.com)

See also:

Tart, Charles T.,Harold E. Puthoff, and Russell Targ, eds. *Mind At Large:Institute of Electrical and Electronic Engineers Symposia On The nature of Extrasensory perception.* New York, Praeger, 1979. Revised and Updated IEEE Symposia papers, 1976-1977 Print

[2] Whitson, T.W., D.N. Bogart, J. Palmer, and C.T. Tart. "Preliminary experiments in group Remote viewing." *Proceedings of the IEEE.* 64.10(Oct. 1976): 1550- 51. Print.

[3] Targ , R., and H.E. Puthoff. *Mind-Reach: Scientists Look at Psychic Abilities (Studies in Consciousness).* New York: Delacorte, 1977. Print.

[4] Puthoff, H.E. and Russell Targ. "Information Transmission under Conditions of Sensory Shielding." *Nature* 251 (1974): 602-07 Print

[5] Atwater, F. Holmes. *Captain of My Ship, Master of My Soul.:Living With Guidance.* Charlottesville: Hampton Roads, 2001. Print.

[6] "Remote Viewing History, Facts, Names, Places and Events." *Psi Tech News*. Psi Tech Inc. 12 Apr 2009. Web <http://www.psitech.net/news/therealremoteviewinghistory.htm>

[7] Puthoff, Harold E. "CIA-Initiated Remote Viewing Program at Stanford Research Institute." *Journal of Scientific Exploration.* 12,1(Spring 1996):Article 3. Print.

[8] Targ , Russell. "Remote Viewing at Stanford Research Institute in the 1970s: A Memoir." *Journal of Scientific Exploration* 10,1(1996): 77-88. Print.

[9] May, Edwin C. "The American Institutes for Research Review of the Department of Defense's STAR GATE Program: A Commentary." *The Journal of Parapsychology*. 60(March 1996): 3-23. Print

[10] Targ, Russell. *Limitless Mind: A Guide to Remote Viewing. And Transformation of Consciousness* Novato: New World Library, 2004. Print

[11] Rauscher, E.A. And Targ, R. "The speed of thought: Investigations of a complex space-time metric to describe psychic phenomena", *Journal of Scientific. Exploration*, 15,(2001) 331. Print

[12] Schulz, Mona Lisa. *Awakening Intuition: Using Your Mind-Body Network for Insight and Healing*. New York:Harmony, 1998. Print

[13] ________. *The Intuitive Advisor: A Psychic Doctor Teaches You How to Solve Your Most Pressing Health Problems*. Hay House, 2009. Print

[14] "Remote viewing" wikipedia.org. 12 Mar 2009. Web <http://en.wikipedia.org/wiki/remote_viewing>

II The Trial Metaphor

The conflict of belief systems of believers and nonbelievers in remote viewing is reminiscent of the Scopes Trial which took place during 1925 in Tennessee. Scopes broke the law of the time by teaching evolution. A Tennessee State law said it was illegal to teach anything that disagreed with the story of Divine Creation of man as taught in the Bible. Since the conflict between the literal teachings of the Bible and science were such hot issues of the time, the case attracted national attention. Religious bigots ran wild. They preached that children who were taught evolution would be dammed to hell because they no longer believed God's word. They accused Scopes of being the devil's servant. Townspeople had bonfire rallies and parades protesting the corruption of children's minds and souls.

Two of the most famous lawyers of the time came to the small Tennessee town to try the case. The trial was reported on the front pages of newspapers throughout the country. Scopes was convicted. However, the verdict was thrown out on appeal.

The trial was a conflict[15] between the ideas of fundamentalism and modernism, a contest between the dogmatism that the Bible was literally true and the beliefs that scientific findings can be true.

Today, we have a similar situation with people's beliefs about remote viewing and ESP. If you tell some people, even scientist with advanced degrees, about remote viewing they will run the Bigots Protocol on you. I have found two kinds of fundamentalism that oppose the belief of remote viewing and ESP: scientific fundamentalism and a philisophical fundamentalism related to quantum mechanics.

Scientific Fundamentalism

Scientific fundamentalism is in the popular dogmatic belief that nothing in science is valid (except possibly quantum mechanics) that was not taught a half century ago and agrees with Newton, Maxwell and Einstein and other scientist within that bracket. A person could hold these beliefs if, like George, they were awarded their PhD's in mathematics and physics before the late part of the twentieth century. A lay person could hold these if their scientific training was limited to high school or the general education science courses in college.

Modernism in the analogy pertains to the scientific basis for remote viewing that has been published in the last decade.

So, on one side of the conflict are those who learned their science in the last century whose belief systems are set in the scientific *Bibles* they learned from. That education didn't allow any psychic phenomena. They will run the Bigots Protocol whenever anyone even mentions psychic phenomena. In most scientific circles, a person is discredited if they even talk about psychic phenomena. I know engineers and scientists who get very angry upon the mere allusion to psychic phenomena. According to one of the most effective remote viewers[16], the intelligence remote viewing program was killed by military officers who ran the Bigots Protocol to rid the military of the heresy.

On the other side of the conflict, there are those who believe in psychic phenomena. Most, like myself, base that belief because we have had personal psychic experiences. It also includes competent scientists and mathematicians such as Hal Puthoff, Russell Targ, and Elizabeth Rauscher who have provided the new scientific paradigm for remote viewing.

This section is in the form a new Scopes trial might take with issues being the dogmatism of obsolete scientific views versus the scientific validity of remote viewing. I will present expert witnesses whose book covers or papers are shown. I will, however, summarize in my words, not theirs, what their books or papers say. I do not presume to be the expert in all these areas. I am only the interpreter in much the same way a lawyer interprets information in a trial.

You, the reader can be the jury. I will present both cases that:

> Remote viewing and ESP are against all scientific principles, or
>
> remote viewing and ESP are real and scientifically verifiable.

The following "Levels of Scientific Proof (Belief)" is part of what might be a set of instructions to a jury.

You are to assign true or false to the claim that remote viewing is scientifically valid and a level of confidence in that verdict.

Levels of Scientific Proof (Belief)

Because I come from the stringently scientifically disciplined commercial communication satellite designer background, I have to qualify the levels of proof of the claim.

George, the respected scientist I described before, has no room in his belief system for anything that does not fit his idea of being scientific. He is paid to operate in a belief system of "Absolutely no conceivable doubt."

Absolutely No conceivable Doubt

In the communication satellite scientific culture in the company where I worked, we would not build and certainly not launch a satellite unless it met the following criteria:

> ***Absolutely no conceivable doubt*** exists that all technical issues had been resolved to the highest scientific experimental standards.

We would test the satellite under all launch and space environmental conditions anticipated before we launched a satellite. Tens or hundreds of millions of dollars were at risk on the satellites working as designed.

In that culture, people like my colleague George, couldn't even conceive of considering a flaky ideas like remote viewing or ESP.

For someone to be convicted in a court of law under the criteria, *Absolutely no conceivable doubt*, the crime would perhaps have to be perpetrated in front of many witnesses, be photographed by newspapers, and appear on at least two TV networks. The crime couldn't be a high profile case like the Kennedy Assassination because people would construct conspiracy theories. The perpetuator would also have to be poor so he could not hire expensive lawyers to raise public doubt like many of today's high profile cases. Everyone would have to believe the perpetrator was absolutely guilty. That's why courts do not use this criterion.

Beyond a Reasonable Doubt

Here, I borrow a legal definition from Criminal Law used to instruct juries as a criterion for conviction.

Most criminal cases do not have the level of evidence described in the foregoing example, so the lesser proof of *Beyond a Reasonable Doubt* is used.

The best peer-reviewed scientific journals require this same level of proof or belief. It is acknowledged that later experimental evidence may make a paper obsolete.

Preponderance of Evidence

Civil cases have this lesser demanding requirement of proof. The persons making the judgment must believe that there is a high probability that the evidence proves the case.

In scientific circles, this is the realm of papers that are based on statistical evidence. The author has gathered enough statistical evidence that they believe their conclusions are valid. One always has to be skeptical of such statistical evidence. The medical world is exposed to many studies proving a *preponderance of evidence* with data skewed for commercial reasons.

Probable Cause

In the legal world, this level of proof requires material facts that support a belief that a crime or violation has been committed. *Probable cause* can be supported by anecdotal evidence or an incomplete case of solid evidence. *Probable cause* is justification for further investigation.

In the scientific world, *probable cause* also is an indicator that further research could be justified. Part of the evidence may be strong with some "weak links" needing to be investigated.

A good example, and a favorite of Carl Sagan, is UFO sightings. There are millions of reports of UFO sightings, many by highly credible people. Blue-ribbon committees of prestigious panels have investigated UFO sightings and concluded that whatever UFOs are, they are "not an apparent national security threat." Therefore, they can be ignored.

True by Analogy

Here an idea is taken from one belief system where it is considered true and used to prove something in another belief system.

Urban Myth

Many people believe things that are really urban myths propagated by hearsay. For example, cats shouldn't wear collars because they will hang themselves in trees. That myth was probably originated by a guy that had his eye shot out by a B-B-gun.

Many Folk Medicine remedies and New Age practices are at this level of proof.

Not True At All

This is the zero point on the belief scale. There is no set of data that support this proposition. It is outside of a person's belief system.

---

[15] Willard B. Gatewood, Jr, *Controversy in the Twenties; Fundamentalism, Modernism, and Evolution.* Nashville: Vanderbilt U. P, 1969. ix + 459. Print

[16] McMoneagle, Joseph. *Memoirs of a Psychic Spy:.* Charlottsville: Hampton Roads Pub, 2006. Print

III The Scientific Case Against The Validity of Remote Viewing

In this section, I will follow the trial metaphor. I first make the case that the claims of remote viewing violate all scientific principles.

I will pretend I am the attorney representing the case against remote viewing. The following is my brief.

No Physiologic Mechanism to Support Remote Viewing Or ESP

> ***There are no known physical structures that can be associated with remote viewing, ESP, or any form of mind-to-mind communication.***

First, there is no physiological mechanism for remote viewing or ESP. Neurologists have never found any structures or functions that can be associated with remote viewing or ESP. Specifically, there have been no transmitting and receiving structures identified. Functional MRI studies by neurologists have never identified any part of the brain that is involved with any ESP activity.

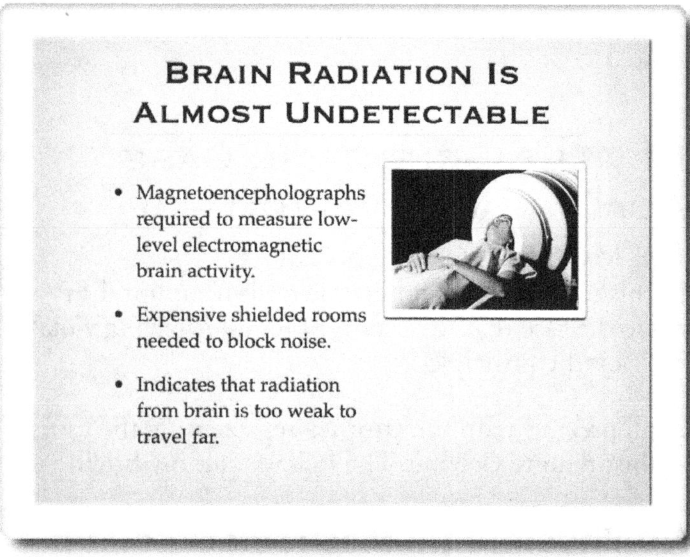

Second, the entire electrical energy created in the brain by nervous activity is too small to transmit information any distance. The above picture shows a research magnetoencephlograph which is the most sensitive instrument for brain research. It is so sensitive it will pick up minute electromagnetic emanations from the earth. It must be located in a state-of-the-art shielded room. Even then, the brain signals are so weak they are hard to detect. In the shown magnetoencephlograph, the magnetic sensors are placed close to the head because the signals are too weak to be detected at any distance.

Everyone that owns a cell phone knows that the farther away from a cell tower you get, the fewer bars or the weaker the signal. When you drive in the country, far from the city stations, your car radio doesn't receive as well. If your head were radiating a radio signal generated by neurons, the power would not be enough to carry farther than-a fraction of a mile at the most. It certainly couldn't carry around the world from Australia. Radio waves would need radio frequency generators, antennas, and receivers, none of which have been found in human physiology.

Third, no substantial scientific evidence exists that, even at short ranges, human beings communicate electromagnetically. That's why we talk and buy cell phones.

Remote Viewing Or ESP Cannot Be Explained By Conventional Physics

> ***If one examines the realms of conventional physics, there is none that explain how there can be long-range remote viewing or ESP.***
>
> ***If remote viewing and ESP exist, they are outside the laws of physics.***

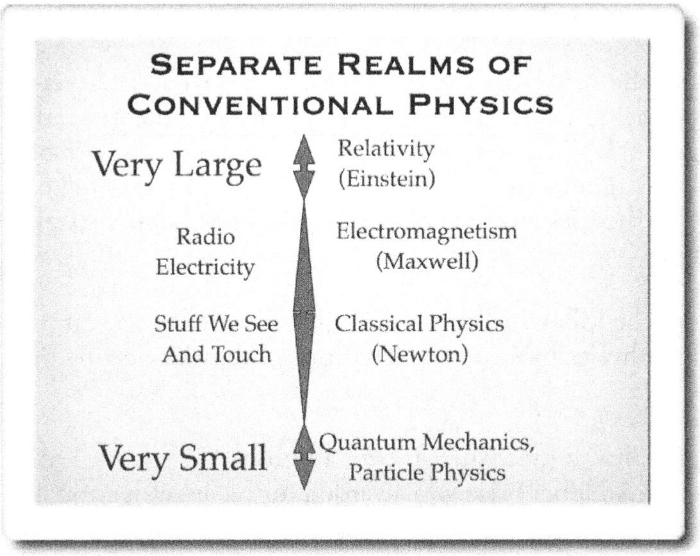

The separate realms of physics are shown in the above figure arranged from the very small to the very large.

Quantum Mechanics and Particle Physics deal with submicroscopic phenomena and have a set of theories and mathematics for dealing with that scale.

On a larger scale, we have phenomena of common experience–stuff we see hear and touch. We know the basics of classical physics intuitively. We use them in walking, throwing things, driving a car, and innumerable other activities in our day-to-day life. The basic mathematics for classical physics are taught in high school.

Then, we have the invisible radio waves permeating our environments and the electricity we can't see in our wires. These phenomena are explained by Maxwell's four equations, each less than an inch long but only understandable by people who have had four years of college mathematics.

At the very large end of the scale, we have the gravitational phenomena explained by Einstein's theories. Although Einstein's theories are explainable to lay persons in diagrams and animations, the full understanding of his theories requires very specialized postgraduate training.

In the following, it is shown that none of these realms of physics have any explanation for remote viewing or ESP.

Quantum mechanics is over a hundred years old. Einstein's Nobel Prize was awarded for his investigation of the photoelectric effect. Some call him the father of quantum mechanics, although he later argued against the theories. Quantum mechanics didn't become its own field until the 1920s when scientists had to invent something to explain phenomena that could not be explained by conventional physics.

In quantum mechanics, things do not happen mechanically as they do in classical Newtonian physics. Nothing conforms to a person's everyday experience. Everything happens probabilistically until you look or make a measurement. Then, it behaves deterministically. This realm of physics is a first for science. It is the first time scientists have recognized something beyond normal senses. Things are indeterminate and statistical in quantum mechanics.

Schrodinger, one of the inventors of quantum mechanics, gave us his famous thought experiment about the cat in the box. He intended the thought experiment to point out the absurdity of indeterminacy in quantum theory: the idea that all probabilities exist until a measurement is made. In the thought experiment, a cat is placed in a box with radioactive atoms and a Geiger counter that will trigger a mechanism to kill the cat when the atoms decay. The box is closed and then again opened when the probability of an atom decaying (and killing the cat) is 50%. The cat is 50% alive and 50% dead just before you open the box. The act of opening the box to make the "measurement" causes the cat to be either dead or alive. Until that time the state of the cat is indeterminate, 50% alive, and 50% dead.

It turns out that Schrodinger's cat is still quite alive. Physicists are still debating the philosophic and mathematical issues of the thought experiment in books published today.

Learning the belief system of quantum mechanics requires undoing the belief systems of classical physics and relativity. It is somewhat like being a member of an extreme religious sect and them being taken to a deprogrammer to strip away all your beliefs. Many physicists are critical of quantum mechanics, particularly when the approach is reductionist. That path leads to string theory.

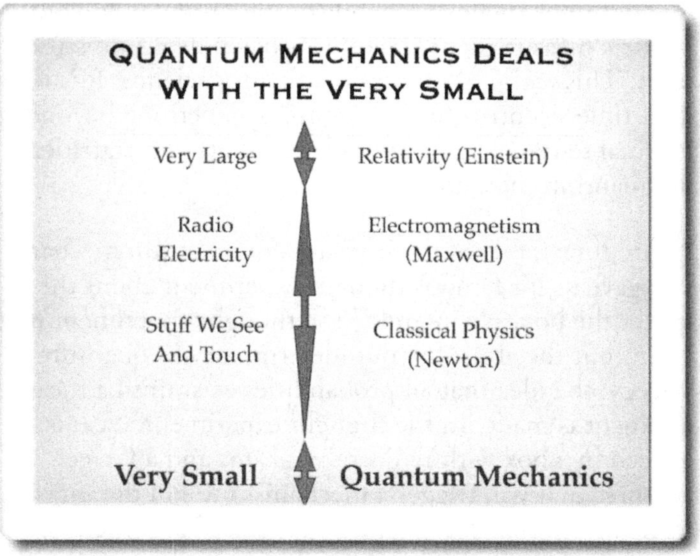

Quantum mechanics only deals with the very small. Quantum mechanics is the deterministic law of small things: atoms, subatomic particles, and molecular structures such as computer chips or some microscopic biologic cells.

Quantum mechanics is a system of mathematics that is appropriate to microscopic and submicroscopic phenomena. It is supported by a vast body of experimental data. It pertains only to things so small most people will never see them.

Quantum mechanics has led to many Nobel Prizes and inventions, including solid-state electronics which have had an immeasurable impact on the world.

Quantum mechanics is a part of conventional physics. It neither predicts nor allows remote viewing or ESP. I know of no literature in conventional physics journals written by people who teach and research quantum mechanics that indicates remote viewing or ESP are predicted by quantum mechanics.

However, the measurement of the phenomena associated with entanglement have been inappropriately appropriated into metaphysical or philisophical speculations. I will deal with those speculations is a separate trial brief in a following section.

String Theory

The current belief system of many theoretical physicists is something called string theory. It came into vogue in the last decade of the twentieth century. It is a Theory of Everything, a unification of all physics.

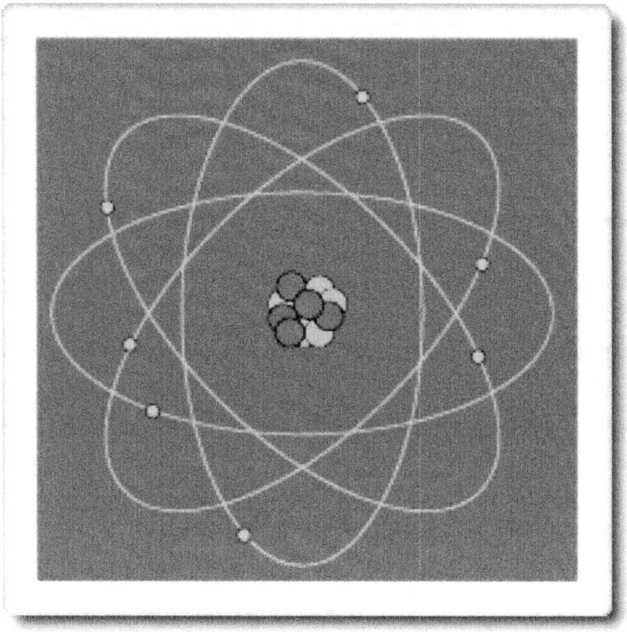

A half century ago, the common belief system was that atoms were nucleuses of protons and neutrons orbited by electrons. This was called the Bohr Model of the atom. Although quantum physicists held the belief that electrons were probabilistic clouds, the Bohr Model of atoms persisted in general high school chemistry and physics textbooks.

Research led to another belief system about the composition of matter. In the 1960's, particle physics experimental data indicated that there were smaller particles. The theory of quarks came into being. In that era, it was believed that protons and neutrons were made of smaller particles called quarks. Current physics texts present the modern evolution of that theory. Physicists now refer to that belief system as The Standard Model. It explains most of the interactions of particles observed in high energy particle colliders.

During this era, another phenomenon came into being that I find the most amazing of all. Governments, particularly European Governments, believed spending billions of dollars on building higher and higher energy particle colliders was important. An eight billion dollar Large Hadron Collider at a CERN laboratory near Geneva is now in operation. The general question these projects hope to answer is:

"What unsuspected structure might exist at the next energy scale?"

Members of Congress or their counterparts in other countries must have the belief system that the "next energy scale" will bring a expenditure of billions of dollars spent in their districts.

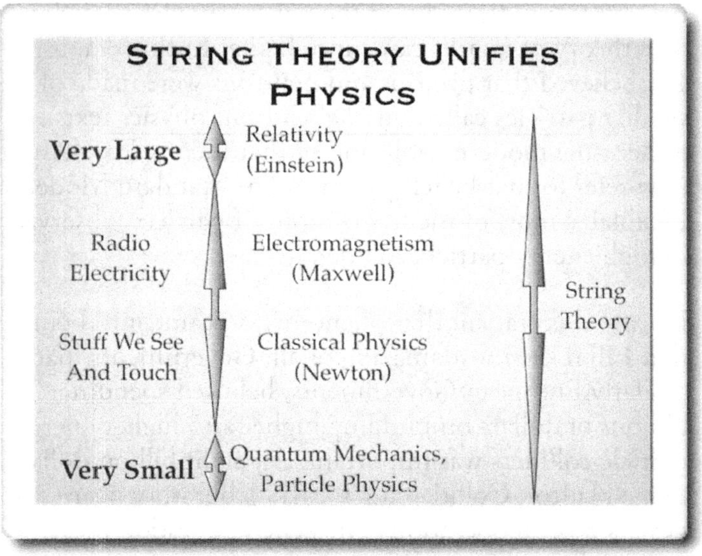

String theory came about to unify the separate realms of physics into a common system. Unification, as the general effort is called, has long been a goal of many theoretical physicists. It is part of the reductionist philosophy that understanding smaller and smaller parts will beget an understanding of the whole.

About mid-century, other scientists began to work in paradigms with a larger number of dimensions. Eventually, in the 1990s, there were five nearly complete competing string theories. Then, a brilliant mathematician, M. Witten showed that all five theories are different ways of thinking about the same thing if the paradigm has eleven dimensions. M-Theory, the ultimate string theory was born.

In the theory, all the fundamental particles are projections of eleven-dimensional vibrating strings into our 4-dimensional space. The exact manner that these strings vibrate in eleven dimensions determines which kind of particle is projected: whether it is a quark, electron, photon, etc. The general equations for string theory fill several pages when written out. By contrast, Maxwell's equations describe all phenomena of light, radio waves and other electromagnetics by four simple equations less than an inch long.

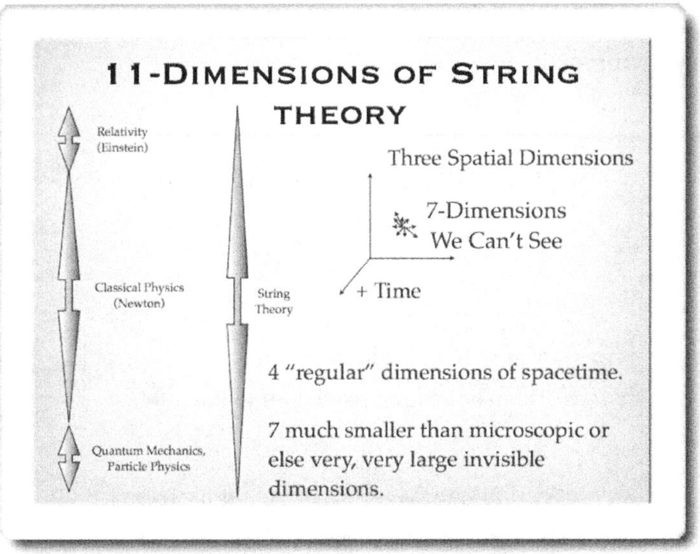

String theory purports to unify the mathematical description of everything from the very small (particle interaction of quantum mechanics) to the very large (bending of light rays of distant stars by planets). But, the idea of 11-dimensions is not in everyone's experience. Where are these seven other dimensions? Mathematicians have a number of conjectures or belief systems on where these other dimensions are. One belief system says the extra dimensions are curled up in much smaller than microscopic 7-dimensional balls. No conceivable experiment or particle collider could ever explore these smaller dimensions. Other mathematicians have belief systems that say there are extra dimensions that are very large membranes or "branes" that are so large we cannot perceive them.

We don't have to try to absorb any of these complicated mathematical concepts or buy any of the best selling books on the subjects. There are simple reasons we can dismiss string theory from being applicable to remote viewing or ESP.

TROUBLE WITH STRING THEORY

- Fundamentals are not supported by any experimental evidence. None!
- The whole thing passed off to the public in TV shows and books-the theory of the "big bang," parallel worlds, alternate realities, universes budding off from ours-is **purely philosophical conjecture.**

The trouble with string theory is it's all conjecture. There is no experimental evidence that supports the theory of strings. String theorists hope that the Large Hadron Collider at a CERN laboratory near Geneva will produce evidence of the existence of particles only predicted by string theory and thereby vindicate the years of work. Until then, string theory is only a purely philosophical conjecture explored by thousands of papers and symposia.

String theory seems to have popular appeal to the general public. Around the hundredth anniversary of Einstein's formulation of his theories of relativity, there were many Discovery Channel and PBS television shows and printed articles about how string theory scientists had finally unravelled the secrets of the universe. Best-selling books told of how relativity, cosmology, quantum mechanics and electromagnetics were brought together by scientists. They could now explain how the "big-bang" happened. They speculated about parallel worlds and alternate realities right next to ours.

I attended a large conference where a prominent scientist gave a lecture where he said string theory now allowed merging of science and spiritual practices or mythologies. In several eastern philosophies, the universe is born over and over again. Scientists say parallel worlds could be budding off from our expanding eleven-dimensional universe. The ultimate goal of meditation is understanding the "oneness of all." String theory is seen as a beautiful way to see the same unity of all things: it exists at the union of science and spirituality.

Soon after the turn of the century, many mathematicians and physicists started to openly express doubt about string theory being anything except a beautiful mathematical theory, unrelated to physical reality.

Certainly, string theory cannot explain remote viewing or ESP. If it is purely an mathematical exercise, then it has no applicability to anything outside of mathematics.

Remote Viewing Violates Laws of Electromagnetism And Relativity

A major problem conservative scientists have with talking about ESP or remote viewing is that these phenomena require the transfer of information over great distances with the miniscule biologically available electromagnetic energy. This transfer is impossible because it violates Maxwell's equations and Einstein's Theory of Relativity.

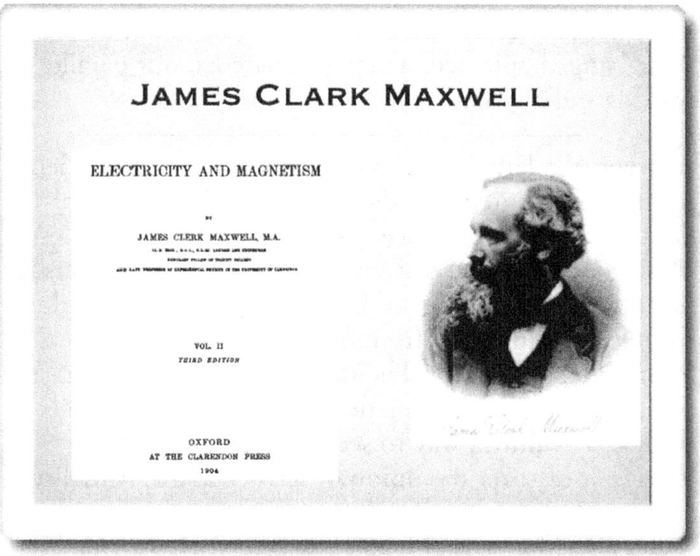

Maxwell's Equations

In classical physics, all fields and flows are described by a set of four equations They are called Maxwell's equations, named after the nineteenth century Scottish mathematical physicist who first formulated them in the 1860s. His books on the subject, shown on the left in the above figure, were still in publication in the early

1900s and are, amazingly, currently available from Google Books.

Maxwell's equations and the later versions of them apply to electricity, radio, and light, as well as quantum electrodynamics (quantum field theory) and parts of relativity theory.

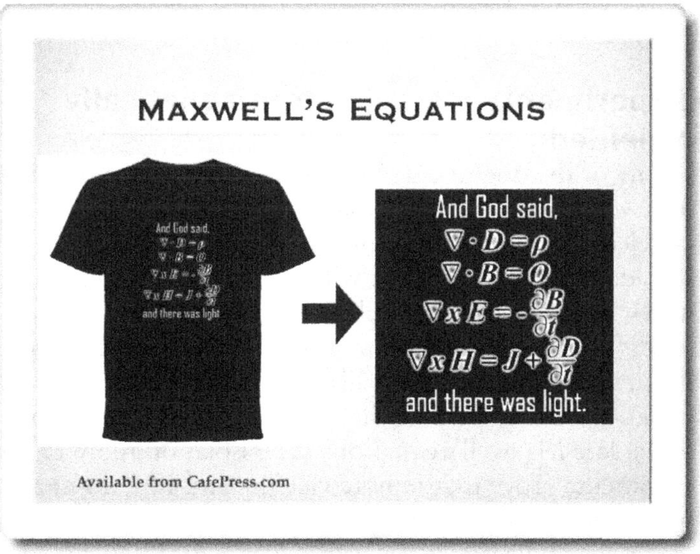

For this brief, it is unnecessary to understand the equations. Above is a joke T-shirt worn by science teachers and student physicists. Maxwell's simple equations on the right describe light and all other forms of electromagnetism. You can't buy any device or piece of electrical or electronic equipment that is not operating according to Maxwell's equations.

The ideas of remote viewing or ESP have no scientific credibility because they don't fit into Maxwell's famous equations. Fundamentally, Maxwell's equations require electromagnetic waves to rapidly decline in amplitude with distance. The claims of remote viewing and ESP are that the communications work as well at great distances as they do locally. That claim is in conflict with Maxwell's equations.

Einstein's Theories of Relativity, in turn, are consistent with Maxwell's equations. There is no room in any of them for remote viewing or ESP.

Experiments Were Electromagnetically Shielded

Many of the documented experiments in remote viewing were conducted while having the remote viewer in an electromagnetic isolation room. These rooms, called Faraday cages, are covered with copper screen and have thick metal doors that latch like walk-in refrigerator doors. These rooms prevent any electromagnetic energy from going through the walls. Again, for any electromagnetic energy to pass through the walls it would have to violate Maxwell's equations. The isolation room experiments prove that remote viewing and ESP do not use electromagnetic energy and therefore cannot have any scientific basis.

Conclusions Regarding Conventional Branches of Physics Supporting ESP or Remote Viewing

The claims of being able to gain knowledge at a distance by remote viewing and ESP are not supported by conventional physics.

Classical Quantum Mechanics pertains to very small phenomena. It's about atoms and molecules and in no way relevant to remote viewing and ESP.

String Theory is a wonderful theory that can unify all fields of conventional physics under one set of mathematics. It has led to very entertaining television programs and cosmology books. However, it is all conjecture, a mathematical exercise: there are absolutely no experimental data to support its theory or any relevance to remote viewing ESP.

Maxwell's equations should be able to explain remote viewing and ESP since they pertain to all electromagnetic communications. However, remote viewing and ESP have distance invariant communications and that violates Maxwell's equations and Einstein's theories of relativity. Therefore, the information cannot be electromagnetically transfered.

Remote viewing and ESP, therefore, have no scientific basis. They are in the category of pseudoscience, superstition, or magic.

The Government's Own Evaluation Of It's Remote Viewing Program Was Highly Negative

The Wikipedia section on remote viewing[17] has been edited by both the proponents and detractors of remote viewing. In the section headed, *AIR evaluation of remote viewing,* the case against the government's program is excellently summarized:

"In 1995, the CIA hired the American Institutes for Research, a perennial intelligence-industry contractor, to perform a retrospective evaluation of the results generated by the remote-viewing program, the Stargate Project. One of the reviewers was Ray Hyman, a long-time critic of psi research, and another was Jessica Utts, who, as a supporter of psi, was chosen to put forward the pro-psi argument. Utts maintained that there had been a statistically significant positive effect, with some subjects scoring 5%-15% above chance. Hyman argued that Utts' conclusion that ESP had been proven to exist, 'is premature, to say the least.'[18]. Hyman said the findings had yet to be replicated independently, and that more investigation would be necessary to 'legitimately claim the existence of paranormal functioning.' Based upon both of their studies, which recommended a higher level of critical research and tighter controls, the CIA terminated the 20 million dollar project in 1995."

According to the official AIR report, there was insufficient evidence of the utility of the intelligence data produced. David Goslin, of the American Institute for Research said, "There's no documented evidence it had any value to the intelligence community."

The Case Against Remote Viewing Rests

This could be the end of the testimony in a remote viewing trial for the side against the belief in remote viewing or ESP. The summary points would be:

> There is no physiological structure to provide remote viewing or ESP.
>
> The claimed distance mind-to-mind capability of remote viewing and ESP violates the laws of all branches of physics including electromagnetism and Einstein's theories of relativity.

The government's own evaluation of the twenty million dollar research program said: "There's no documented evidence it had any value to the intelligence community." The program was terminated.

---

[17] "Remote viewing" Wikipedia. Web, 22 Apr 2009 <http://en.wikipedia.org/wiki/Remote_viewing>.

[18] Hyman, Ray. "Evaluation of a Program on Anomalous Mental Phenomena". *Journal of Society for Scientific Exploration,* 10.1: Article 2.Print

IV The Case For the Validity Of Remote Viewing

Continuing with the trial metaphor, I state the case that the claims of remote viewing and ESP are fully supported by scientific principles and experimental data. However, those scientific principles are not common knowledge to most people since they have only come to fruition in the last few years. Also, they require an uncommon knowledge of mathematics.

As the case is presented here, the advanced mathematical concept of higher dimensions will be present in simplified form. Many people will be uncomfortable with these concepts because they have not had years of exposure to the underlying mathematics. The case will be presented to appeal to the belief that if the experts cited are credible, then the complex mathematic concepts must be credible.

First, we will give you a special knowledge about Einstein's theories of relativity. Fortunately, all you have to know about Relativity, as far as remote viewing and ESP are concerned, is that you don't have to know anything about it. All that one must understand is that Einstein's theories created a horizon. It will be shown that understanding the scientific validity of remote viewing and ESP requires one go beyond that horizon.

Remote Viewing Is Scientifically Explained By Physics When The Mathematics Go Beyond Einstein's Horizon

Einstein's Horizon

To understand Einstein's Horizon one must first look at Einstein's life. His modes of approaching problems and the way he communicated his theories reinforced classical physics' thinking horizon which is limited to three dimensions plus time.

Abbreviated Biography of Einstein.

His Early And School Years

Many biographies have been written about Einstein. My favorite is Walter Isacson's *Einstein*, a wonderful read. Much of the following is my reading between the lines of that book.

Einstein was born in 1879 in Germany. He had early childhood developmental problems. By the age of two he used few words. He would spend time muttering unintelligibly to himself. As he grew older, he would rehearse anything he would to say muttering to himself.

His inability to communicate easily caused him to be somewhat of an outsider at school. He generally hated school because teaching was based on the Germanic style of memorization, repetition, and rote learning in a highly disciplined environment. Because he was different, many other students laughed at him. He would avoid playing with other students and sit muttering to himself.

I asked my wife, a veteran kindergarten teacher, about her experience with unusual children similar to Einstein. She said some kids are unusual and simply "march to a different drummer." Some live in a nerdy, parallel world. They can be very talented but don't fit into the normal classroom. If they are really interested in something, it consumes their attention, and they can devote incredible time to learning about it. If they aren't, they can withdraw into themselves. In the modern school where she taught, a student such as Einstein might be assigned to a "special education" class with other kids who, for many reasons, didn't fit in the normal classroom. Chances are today someone would put him on drugs to try to alleviate his attention disorder.

In spite of his peculiarities, Einstein got good grades. Contrary to myths, Einstein was always excellent in mathematics. He had mastered integral and differential calculus before he was fifteen. He was often the highest student in his math class. However, perhaps because he was smarter than most of his teachers and able to grasp concepts quickly, he was very disruptive in classes. He either was asked to leave or otherwise decided to leave his high school at age 16. He was a high school dropout. If he had stayed in Germany, he would have been drafted into the military at age 17. So, the high school dropout also became a draft dogger and moved to Switzerland to eventually become a Swiss citizen.

In Switzerland, he enrolled at a very different kind of school. This school was based on the ideas of an education reformer of the early nineteenth century,

Johann Heinrich Pestalozzi. Pestalozzi believed education should develop a visual style of thinking. Observation and visualization were primary learning tools. Visual thinking was thought to be one of the most powerful features of the mind, and imagery was the start of all knowledge.

Instead of dealing with words (which Einstein was poor at) students at the school Einstein attended were encouraged to learn through observing things and visualizing relationships. They were free to pursue their own interests and think independently on all subjects. Einstein flourished at this school and graduated second in his class. If it hadn't been for his low marks in languages, particularly French, he would have graduated first.

The visual thinking approach reinforced Einstein's visual style of thinking. Later, as a physicist, he would rely on visual thought experiments such as people riding in trains going near the speed of light and being observed by stationary people on a station platforms. These thought experiments would become the hallmark of Einstein's theories.

I suspect that if Einstein was graduating from high school today, he might have skipped going to a university and become one of those brilliant minds that program computers and become millionaires by designing video games.

At The University

Einstein enrolled in Zurich Polytechnic. It was a small school of 71 students with a solid reputation for science and engineering. A section of the curriculum was designed to produce teachers for teaching high school mathematics and physics. Einstein was one of eleven students enrolled in that freshman section.

During his first couple of years, he was an enthusiastic student. His instructors initially loved his enthusiasm. Then, he began to get bored and skip classes. A friend of his, Marcel Grossmann, took elaborate classroom notes. Einstein could quickly study the notes and pass the tests. At the end of two years, students had to pass an oral exam. Einstein got the highest scores in math and physics.

During his next two years, Einstein became disruptive in the classes he did attend and fell out of favor with all his instructors. His primary physics instructor was Heinrich Weber. Einstein cut most of his lectures and once complained that Weber's lectures were fifty years out-of-date. Herman Minkowski, another mathematics teacher, also didn't get along with Einstein because Einstein also cut most of his classes. Again, Einstein relied on Marcel Grossmann's notes to pass the exams. Einstein spent most of his time on the extracurricular study of contemporary papers reporting on the frontiers of physics. Einstein's grades fell and, at graduation, he was fourth-of-four passing candidates for a degree in physics.

At The Swiss Patent Office

After graduation, Einstein applied for many positions teaching math and physics. However, Professor Weber, would only give negative recommendations to potential employers. Einstein was unemployed for two years. Finally, as the result of a friend's family connections, he got a job as a patent examiner at the Swiss Patent Office.

While at the Patent Office, Einstein had ample time to continue his studies in physics and write scientific papers. In 1905 he published four of his most important papers. He wrote a paper that said light was a quanta of energy. That paper established the photoelectric effect and started the idea of quantum physics. A paper on the random movement of very small objects was important in supporting atomic theory. He then published his radical theory of special relativity. That was followed by a paper about the mass-energy equivalence that produced the familiar relationship $E=Mc^2$.

In 1906 he submitted his paper on relativity as a doctoral thesis to the University of Zurich. It was turned down as too radical. He submitted the previously written paper on random movement of very small objects and received his doctorate.

Initially, these four papers didn't get Einstein much recognition. However, later physicists would refer to 1905 as Einstein's miracle year. These four papers reshaped some of the fundamental space and time concepts of physics that had existed for two hundred years. Einstein continued working at the Patent Office until 1911 when his recognition was great enough to get him an academic job.

Einstein's Later Career

Although Einstein wrote over three hundred papers in his career, his four 1905 papers and the 1915 paper on General Relativity presented the concept for which he is most famous. Einstein's Special Relativity is "special" because it is a special (not general) case. It pertains only to uniform motion, like traveling in straight lines. The thought pictures he used in describing the theory, included trains traveling on straight tracks at constant speed while being observed by stationary people. His later theory of General Relativity included both uniform motion and acceleration. The thought pictures could include planets orbiting around a sun or elevators accelerating through space in linear or circular trajectories.

Einstein won the 1921 Nobel Prize in Physics for his work on the photoelectric effect described in his 1905 paper.

Einstein's fame continued to rise. His theory of General Relativity became the subject of experiments by astronomers. Their efforts to prove his theories became front page stories on newspapers. Outside the community of physicists, Einstein became an archetype, synonymous with mastermind, genius and brain. His wild haired appearance added to making him an icon and the superstar of physics. He was also an active Zionist and that organization used him as a poster boy for their movement. That organization publicized, promoted, and exploited his fame. When he came to New York in 1921, his visit was front page news on the newspapers. Reporters solicited his opinions on all sorts of topics. When he spoke at rallies, thousands of cheering fans showed up.

His fame continued beyond his death in 1955. In 1999 his picture on the cover of *Time Magazine* identified him as the "Person of the Century."

In reality, Einstein contributed little beyond his celebrity to physics in the last thirty years of his life. About the time Einstein was receiving the Nobel Prize, the hot new topic in physics became quantum mechanics. Although Einstein is sometimes called the "father of quantum mechanics," because of his theory of light as particles, he never embraced quantum mechanics. Up until 1935 he wrote papers trying to refute ideas of quantum mechanics. Physics passed Einstein by. Einstein spent the rest of his life in the unsuccessful quest for a Unified Field Theory. Physicists both revered him for his accomplishments and dismissed him as irrelevant to the physics of the day. He continued to be famous for being famous and became the icon of the wild haired archetypical genius.

I believe Einstein couldn't move into modern physics because of the horizon of his three dimensional visualizations.

Einstein's Horizon Of Four Dimensions

In the late nineteenth century, philosophers and scientists began considering adding dimensions. In 1884 a Shakespearean scholar, E.A. Abbot published *Flatland, A Romance of Many Dimensions* which toyed with the idea that there may be a fourth spatial dimension and that four-dimensional beings could exist without our knowledge.

Around the start of the century, others postulated the idea that there may be a fourth dimension where all the departed spirits existed. The New York Times frequently had articles about spiritualists and the ideas of spiritualism. Harry Hudini, the famed magician, conducted seances and incorporated the ideas into his stage acts. Spiritualism flourished until the start of World War I. It seemed to disappear from the national consciousness after the war.

Theosophists claim[19] Einstein was a student of theosophy and derived some of his key ideas from those readings. Theosophy was founded by H. P. Blavatsky. In her 1888 book[20] she says:

> "Everything that exists has only a relative, not an absolute reality...."

These ideas might have influenced Einstein in thinking about relativity. Theosophy included ideas that the fourth dimension was the realm of the spiritual.

It is not surprising that Einstein, trained in science, would have thought of ideas of extra dimensions as metaphysics and not been comfortable with the idea as part of physics.

His high school training in Pestalozzi's visual thinking made Einstein think in three dimensions plus time. Up-down, right-left, forward-back, and time were all there was. The rest was metaphysics.

In Einstein's paper[21] on Special Relativity in 1905, he talked only of three dimensions plus time. In 1907 his mathematics Professor, Herman Minkowsky, one of many mathematicians working with the mathematical ideas of higher dimensions, corrected Einstein's "homework" and published a paper[22] in which *time* is also a dimension.

In Einstein's 1913 paper[23] on General Relativity, he addresses four-dimensional Space-Time. Einstein spent the last thirty years of his career trying to develop a unification theory. He failed. Once he tried going from his 4-dimensional space-time paradigm to a 5-dimensional paradigm. He gave it up

Einstein didn't like the ideas of probability that arose in quantum mechanics. He said he didn't believe that God played dice. For instance, he thought that an electron would either be somewhere or not: an electron's location is not a probability. How could he visualize an electron probably being somewhere and probably being somewhere else?

Einstein's Relativity theories deal with light, gravitation, and other large scale phenomena, applicable to planets and galaxies. He was never comfortable with quantum mechanics which only deals with the very small, the atomic level. There, it is difficult to deal with the phenomena by visualization.

Einstein's world of imagination was limited to what he could visualize in time and a three dimensional world. Any dimensions beyond those were beyond his horizon.

That horizon agreed with Newton's and Galileo's. It is the horizon that agrees with our everyday experience and general education. Newton thought space and time were absolute: a meter is always the same thing here or on the moon, a second is the same everywhere, and linear motion will take you in a straight line. Gravity is a fundamental force of nature.

Einstein's General Theory of Relativity upset two hundred years of physics by saying that there is no absolute space or time: matter and energy distort time and space; gravity is caused by the warping of space-time by mass and energy. There are thousand of books available today which explain these concepts. However, it is unnecessary for you to understand them to understand the case regarding the validity of remote viewing and ESP. In Einstein's relativity there are only the four dimensions of space and time. Those defined his horizon.

Our species evolved to perceive everything in Euclidean space. What we see, smell, hear, and feel is what is there in a three-dimensional world. I have never known anyone that claimed to be able to visualize objects in more than three spatial dimensions.

That was also Einstein's horizon.

"Tombstone Knowledge"

I will introduce you to some of the concepts of the graduate-level mathematics and physics that are necessary for understanding that remote viewing and ESPs are scientifically valid. I do this with apologies to mathematicians who will have problems with simplifications of very complex ideas.

I like to think of mathematics and physics as requiring "tombstone knowledge." Mathematicians deal with the life-accomplishments of people by reciting their names. I like to think of these references as "tombstones."

I recently went on a graveyard tour led by Dawn, the local historian of the small town where I live. We started at the far back of the graveyard where the tombstones were of those who founded of the town a hundred and forty years ago. Dawn told us the name of who was buried there, told us what they did and during their lifetime, their families, and what they were noted or stood for. We skipped grave sites of innocuous people. Dawn gave us the town's history in terms of the people and their contributions. We finished at a recent grave of a marine killed in Iraq.

Studying mathematics and physics is like a tour of a graveyard. In the oldest section of the graveyard, you become acquainted with people like Euclid. You learn to say, "Euclidean Geometry" and everyone knows what you mean and you thereby incorporate his body of work in what you are talking about. You learn to say, "Archimedes' Principle" and everyone will know you are talking about why boats float or something about displacement of fluids. More toward the front of the graveyard, we have all the mathematicians and physicists of the eighteenth and nineteenth centuries. Many living scientists and mathematicians have "earned" their tombstone by published papers that are often referenced: their inscription might be "John Doe, mathematician, 1945-...." with the year of death left off. There are many of tombstones that have to be studied today. Here, we learn to say "Boyle's Law" and everyone will have an idea of the equations or principles that will fit in our argument. In college, you might take and entire course or degree about one tombstone. In modern physics, people talk about "Einsteinian Relativity," "Heisenberg's Uncertainty Principle," "Schrodinger's Cat" and "Feynman Diagrams."

Obtaining an advanced degree in physics or mathematics is largely learning what each tombstone means, the underlying concepts and how to talk-the-talk of tombstones. I recently saw a paper on String Theory by one of the leaders in the field which referred to twenty-three tombstones on the first page. Understanding the referenced tombstones probably would have required the understanding of another hundred underlying tombstone concepts.

My colleague, George,who did not like me talking about vibrations, had taken math and physics courses and could reference all the appropriate "tombstones." However, they all assumed the same horizon. He thought of Einstein's relativity theories as the ultimate acceptable description of reality which included Maxwell's equations.

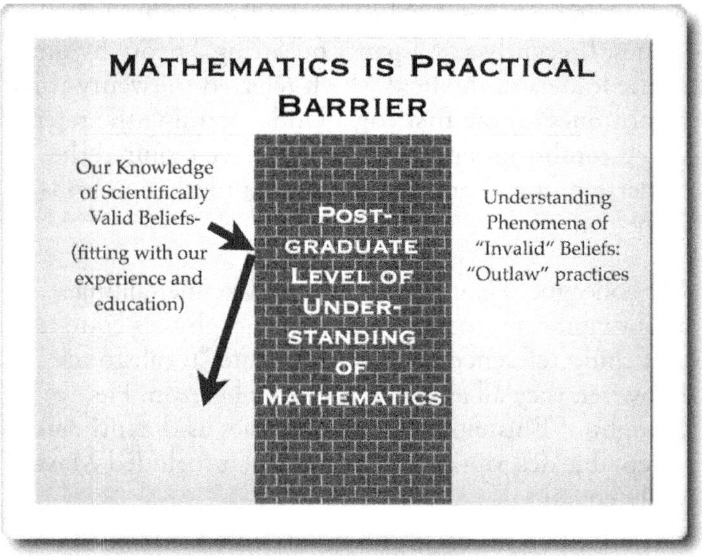

The Basic Barrier of Higher Mathematics

Our instruction in the belief system of what is valid scientifically begins when we are in grade school and high school. We are taught by people with a limited education is Science and we learn the classical form of Science, a belief system that is about a hundred years old. Thinking which leads to the understanding of remote viewing and ESP requires an understanding of higher level mathematics and physics.

By analogy, most of us are taught our basic religious belief system by nuns or Sunday school teachers, not theologians. We learn the rote elements. And then, we learn to disagree with anything that does not fit our knowledge of the rote elements.

Many authors of books on subjects related to quantum mechanics do not have graduate level training in the mathematics of quantum mechanics. They use mystical concepts like "quantum field," "entanglement" or "dark energy" to explain what is beyond the mathematical barrier. Quantum mechanics replaces God as the mystical explanation of the unknown. They are not looking beyond Einstein's Horizon: they are looking toward a "Heaven" where all the mysteries flow from.

As one travels around the internet or attends conferences where ideas on "outlaw" experiences are presented, they can find a huge number of mathematical "theories of everything" couched in esoteric mathematical terms going beyond Einstein's Horizon. One would need advanced degrees in physics and mathematics to competently evaluate these theories. I have studied many of these only to find a fatal flaw in the reasoning at the foundation of the theories. Fortunately, we can stay within mainline physics and get a paradigm that explains remote viewing and ESP. My filter for what is a "mainline" scientific belief and apocrypha is Roger Penrose's book.

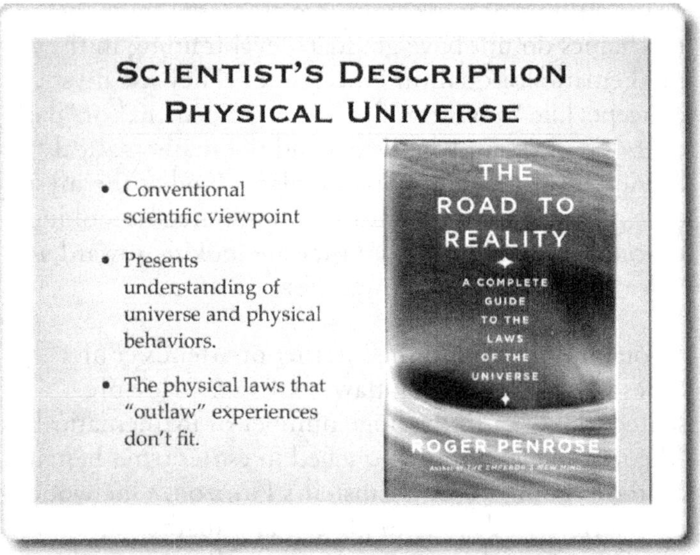

For those of you with some mathematics background who wish to know more about the tombstones I will summarize, I would recommend Roger Penrose's book[24] shown above. Sir Roger Penrose is a retired professor of mathematics at the University of Oxford, and a collaborator with Stephen Hawking on black hole theory. The original London publisher of the book calls it the most important and ambitious work of science for a generation.

He presents a conventional scientific viewpoint of physics in explaining the universe. It covers subjects from quantum physics to Einstein's relativity and introduces all the mathematics that, for the sake or readability, I avoided in this book.

I have made an effort to keep away from referencing many tombstones since my intent is to validate the case for remote viewing and ESP for people whose education did not include all the tombstones. I will try to explain things in an approximate way to communicate basic concepts. I do this with apologies to mathematicians and physicists who have read all the tombstones and people who find Sir Penrose's book an easy read.

Beyond 4-Dimensions

The well-known subjects of quantum mechanics and string theory go beyond Einstein's Horizon. They both use different ideas about dimensions beyond four. These subjects both are considered to follow the laws of physics and are therefore "lawful." Neither of these subjects can describe the ESP experiences of vibrations and remote viewing.

The mathematics must go in a different direction beyond Einstein's Horizon to make ESP experiences of vibrations and remote viewing "lawful." The dimensions must include complex numbers. I will start there.

Complex Numbers

I will avoid mathematical equations. However, it is necessary to review the concepts of imaginary numbers because they are fundamental to the whole idea of going beyond Einstein's Horizon.

I didn't run across complex numbers, numbers made up of real and imaginary parts, until I was an undergraduate in college. Now, they are taught in junior high school. Although they are still a mystery to many–those who "hate math"–they are useful as a mathematical convenience in describing and analyzing things.

A complex number has a real and imaginary part. It is written in the form of:

> $a + ib$ where "a" and "b" are real numbers and "i" is the square root of (-1).

"a" and "b" can be added, subtracted, multiplied, and divided like normal numbers. A complex number can be viewed as a point in two dimensional space generally called the complex plane.

I won't try to give you a course in complex numbers. But, fortunately they are part of most people's belief systems of being acceptable scientifically. Either you have had enough mathematics to be comfortable with equations with complex numbers, or you have heard of them as being part of the acceptable ideas of mathematics

A Theory of Higher Dimensions

Most people, including researchers in other fields such as medicine, are happy with a 4-dimensions. To go beyond Einstein's Horizon one must understand the ideas of higher dimensions.

In this brief, I hope to provide a roadmap to an understanding of higher dimensions which will allow you as well as technically trained people to believe that remote viewing and some ESP experiences are scientifically valid. I wouldn't claim that I am an expert at many subjects to be visited on that road. I wouldn't claim that we are visiting the only relevant tombstones, or that there are not entire subjects that should be addressed. My goal is to connect subjects and create a systematic argument. First, we need to conceptually transport our thinking from a 4-dimensional world to one of higher dimensions.

Need For More Dimensions

In the 4-dimensional physics world, there is no explanation for transfer of information in remote viewing and ESP. Still, people have these experiences. Remote viewing has been scientifically demonstrated and documented. However, many with the 4-dimensional physics belief system still reject the experimental demonstrations.

A pioneer researcher in remote viewing, Russell Targ joined with a brilliant mathematician, Elizabeth Rauscher, and came up with a simple solution: add more dimensions to the physics. In their landmark paper[25], they built on the work of Einstein's math teacher, Herman Minkowski to go beyond Einstein's Horizon. I will describe their theory in much greater detail in a subsequent chapter. First, however, I will stop at Minkowsky's "tombstone" and discuss how Minkowsky's work provides a bridge of belief between the 4-dimensional belief world and one in which "outlaw experiences" such as remote viewing and ESP are "lawful."

Minkowski Space

Herman Minkowski was born in Lithuania in 1864 and became Einstein's mathematics teacher at Zurich

Polytechnic. After Einstein's graduation, he was familiar with Einstein's work from their contact in the post graduate work Einstein was doing at Zurich Polytechnic. After Einstein published his paper on Special Relativity, few people were aware of it. There were some problems with reconciling Einstein's Special Relativity with all the independent, similar work going on in mathematics around 1905. Minkowsky was one of the first people to grasp the importance of Einstein's theory in 1907, two years after it was published. He proceeded to expand the mathematics around the theory and sell it to other mathematicians. In 1908, Minkowsky advocated considering *time* as a fourth dimension.

Some people say that he nearly hijacked development of the ideas of relativity because he thought the ideas were too important to be left to physicists such as Einstein. He is quoted as saying about relativity to another physicist, "It came as a tremendous surprise, for in his student days, Einstein had been a lazy dog. He never bothered about mathematics at all."

In Minkowsky's 4- dimensional space definition, there is something that creates difficulty for people like Einstein that like to visualize geometries. The fourth dimension, *time*, is an (mathematically) imaginary dimension that is measured in terms of the square root of (-1).

Einstein initially rejected Minkowsky's mathematic approach as too complicated. I conjecture that he rejected it because it was beyond his horizon.

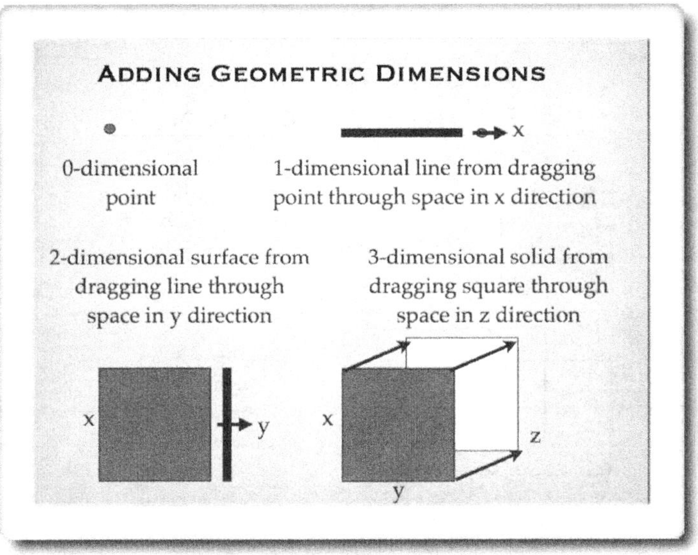

The basic idea of dimensions is shown in the above diagram. Mathematically, a point in space can have no dimensions. Drag the point through space in one direction, call it the *x* direction, and it becomes a straight line of one-dimension. Then, drag the line in a direction perpendicular to the *x* direction, call it the *y* direction, and you produce a two-dimensional rectangle. Drag the rectangle in a direction perpendicular to the two-dimensional plane, call it the *z* direction, and you have a three-dimensional box. This is all easy to visualize within Einstein's Horizon.

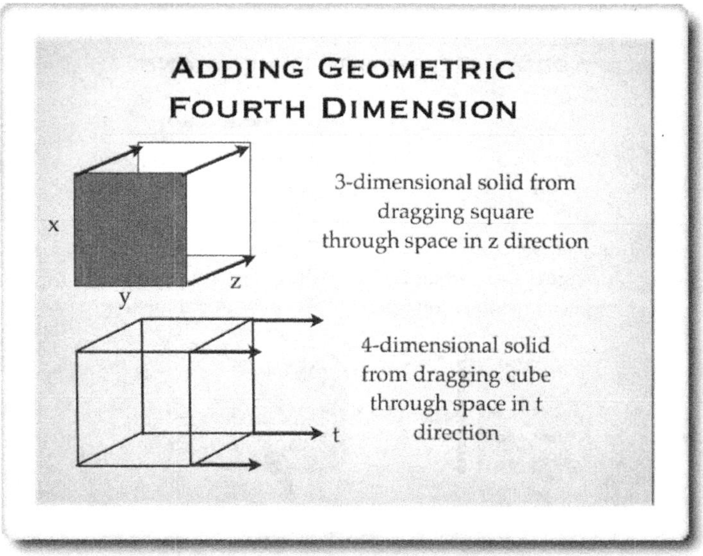

Next, add time as a dimension. If a bullet was shot through the box, we could represent its path as a line with time ticks along the path. Then, we would have visualized four-dimensions. That is also easy. However, four-dimensions are really created by dragging the three-dimensional box through a time dimension. That is harder to visualize. But, one can be comfortable with the time-ticked line in a three dimensional space if you have worked with geometry enough to understand that it is a three-dimensional representation of slices through a four-dimensional space. All this sort of visualization is on the edge of what humans can do, the edge of Einstein's Horizon.

Mathematicians are not bound by what can be visualized. They work with equations and are perfectly comfortable defining all sorts of abstract spaces of any number of dimensions. String theory, initially used twenty-one dimensions, although no one had any idea how to visualize them.

When Minkowsky call *time* another dimension in extending Einstein's relativity, he threw in a twist. To make his mathematics agree with other mathematician's work in the area of relativity he had to make *time* an imaginary dimension: in his equations *t* was multiplied by "*i*" which as explained in a preceding section, is the imaginary square root of (-1).

This is beyond my horizon. I can't visualize "imaginary time." What is it?

I speculate that when Einstein saw Minkowsky's equations which dealt with imaginary time, it was beyond Einstein's horizon. That's why he rejected Minkowsky's approach as too difficult to understand.

Eventually, he adopted Minkowsky's 4-dimensions of space-time. That shift was important in allowing him to eventually come up with General Relativity. However, his horizon was limited to these four dimensions.

Since then, people have talked about 4-dimensional space-time as I have in this book. It is also commonly called Minkowsky space. There are diagrams[26] presenting visualizations of Minkowsky space. However, they are somewhat difficult to understand and unnecessary to the reasoning I am following.

Quantum mechanics moved beyond 4-dimensional spacetime and Einstein refused to go there and accept the theories. String theory, which came after Einstein's time, moved beyond 4-dimensional spacetime. Neither of those can explain remote viewing and ESP. Another extension beyond 4-dimensional spacetime is needed. To convince scientists whose thinking is also limited to Einstein's Horizon, the extension needs to done in a way to be compatible with Quantum Mechanics, Classical Physics, Electromagnetism, and Relativity.

Moving Beyond Einstein's Horizon

> ***Honorable, peer-reviewed publications in conventional physics can provide a scientifically valid explanation of how remote viewing and ESP can work over large distances.***

Those Who Have Gone Beyond Einstein's Horizon

As we have mentioned before, string theory is beyond Einstein's Horizon. The media tells us that string theory is one of the ultimate quests in physics. Edward Witten, a theoretical physicist at the Institute for Advanced Study at Princeton, coalesced competing string theories into something he called M-theory. Some of his peers rank him as the greatest living mathematician and sometimes greater than Einstein or even Newton. He was on Time Magazine's list of 100 most influential people of 2004. He, like a great number of other mathematicians seem to have gone in a direction that can't explain "outlaw" experiences. Some critics, who point to the lack of experimental evidence to support string or M-theory, quip that the theories are no more than a very complicated version of sudoko.

Elizabeth Rauscher is a mathematician and physicist that didn't go down the *rabbit hole* after string theory. Some of her peers call her a female Einstein. However, unlike Einstein whose major accomplishments were only early in his career, she has had a long career and her accomplishments continue to grow. She had published more than 260 papers over more than forty years, many of which were on subjects off the mainline of physics. Her 2008 paper[27] with Richard Amaroso on *Relativistic Physics in Complex Minkowski Space* is her latest paper on the 8-space that goes beyond Einstein's Horizon. Her papers with Russell Targ[28,29] specifically relate to how the 8-space can be used to give a scientific explanation for remote viewing, ESP and precognition.

Why Pick An 8-Dimensional Minkowski Space?

Most people who do not read a lot about physics may not realize how many different models and theories there are about reality: Loop Quantum Gravity, lattice approaches, Euclidian Quantum Gravity and Twistor Theory, to mention a few described in the peer-reviewed literature of physics. On the internet, you can also find hundreds or maybe thousands of self-proclaimed quantum physicists and other assorted people who have read physics and made up their own theory of how things work. Many have invented mysterious energies to explain ESP phenomena.

The 8-dimensional complex Minkowski space has been shown to be consistent with our present understanding of the equations of Newton, Einstein, electromagnetic theory and the foundations of quantum mechanics, It is advocated by credential academics in peer-reviewed journals. It has been under consideration for explaining "outlaw" phenomena for over two decades.

In 1983, Elizabeth Rauscher[30] extended Maxwell's equations into complex 8-dimensional Minkowsky space. In 2002, Elizabeth Rauscher and Richard Amoroso[31] further described Maxwell's equations in terms of complex 8-dimensional Minkowsky space. In 2006, they created a more extended model[32] that shows that standard quantum theory is a linear approximation to a higher dimensional complex space.

This 8-dimensional complex Minkowski space seems to encompass the major theories of "lawful" physics:

- Newtonian mechanics that we all learned in high school physics,
- Maxwell's equations that describe electromagnetics,
- Einstein's Relativity, and
- Standard Quantum Mechanics.

If remote viewing and ESP phenomena can be explained in terms of 8-dimensional complex Minkowski space, they can be considered "lawful." I will first describe the 8-dimensional Minkowsky Space in metaphorical terms.

The General Idea of Eight-Dimensional Space

As previously mentioned, most of our species cannot visualize more than three or four dimensions–not even Einstein himself. My personal approach to dealing with higher dimensional geometries is to visualize a problem in terms of our ordinary three dimensions and then, as an act of faith and intuition, believe that something analogous is happening in the higher dimensions.

There are two ideas of distance in the 8-dimensional realm. The first is the *mechanical distance* which is the distance concept of common thinking. *Mechanical distance* measures the separation of things in our 4-dimensional world, that we can see. The second kind of distance is *Apparent distance* which in a 4-dimensional world is a matter of illusion or visualization. If I ask you to visualize the Eiffel Tower, the *apparent distance* between you and the Eiffel Tower is decreased to whatever viewpoint you choose in your visualization. Unless you are in Europe, the *mechanical distance* is thousands of miles.

Illusions can be created which confuse your perception of distance. In a theme park I visited, they had a house that was haunted by illusions of spooks. The spooks had apparent distances that fooled me. Here is how they might have worked. On the spook-ride, we were in little cars riding on a track. One of the rooms we went into was the Spook Library.

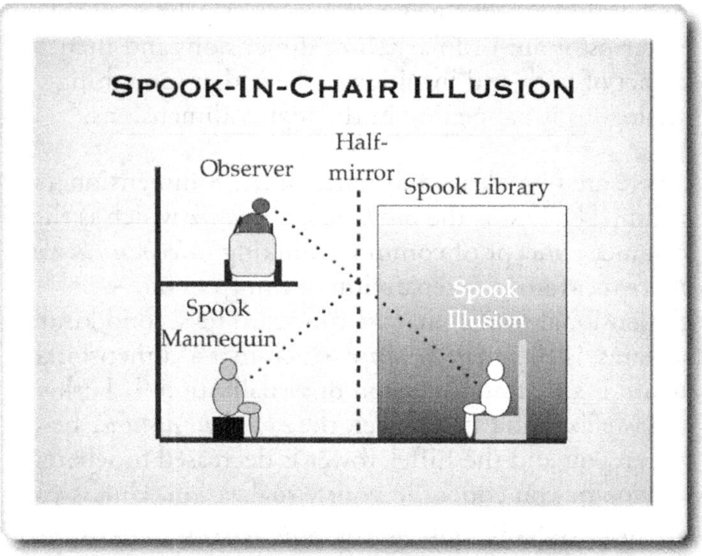

In this illusion, the car and the observer travel past the Spook Library on a balcony in a big room. As I looked down into the dimly lit room, I saw a normal library with bookshelves and a carpet and a chair. However, there was a semitransparent spook sitting in the chair. I could see the chair right through him. As we moved through the room on the balcony I could see that he was three dimensional. He waved his arm at us.

This above figure shows how the illusion was created. There was an invisible half-transparent mirror between the observer and the Spook Library. An animated mannequin painted in fluorescent colors was under the balcony. I saw the reflected image of the mannequin superimposed on the chair.

The *mechanical distance* between the mannequin and the chair was the distance from the mannequin to the mirror plus the distance from the mirror to the chair.

The *apparent distance* between the mannequin and the chair was zero.

The zero apparent distance illusion only works when the chair, mannequin, and mirror are a certain set of distances from each other. If I drew a set of *x, y,* and *z* coordinates in the room, there would only one set of *x, y,* and *z's* that would work to allow the illusion. If the mannequin were too close or too far from the mirror the *apparent distance* would not have been zero.

In another part of the theme park ride there was a sign that said, "beware of hitchhikers." That is the next illusion.

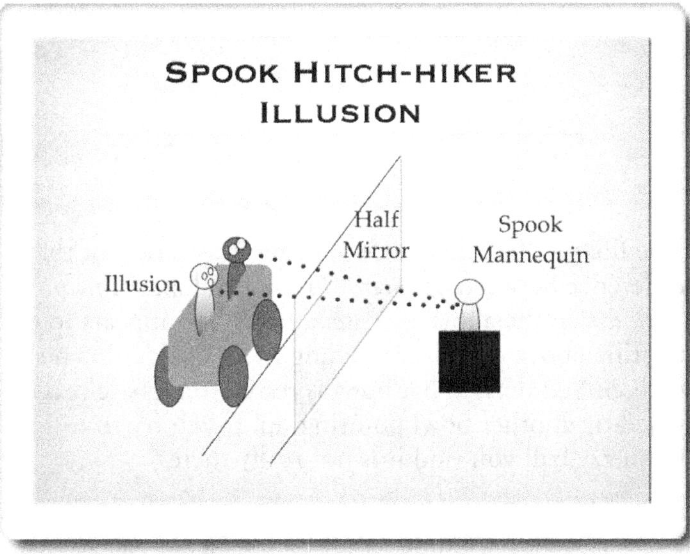

In this illusion, when I looked into the mirror I saw a transparent spook sitting next to me in the car. The apparent distance was one foot to my right. Again, the illusion was created using a half-transparent mirror with the spook mannequin behind the mirror. The *mechanical distance* was that from me through the mirror to the mannequin, perhaps twelve feet. The *apparent distance* was one foot.

FLOATING BALL ILLUSION

The floating ball illusion is another illustration of the difference between *apparent distance* and *mechanical distance*. In the above picture, a fuzzy ball appears to be floating above the silver opening in an upside-down six inch curved bowl. That bowl is resting on a base provided by another bowl pointing up. If you try to touch the fuzzy ball you find it is not really there.

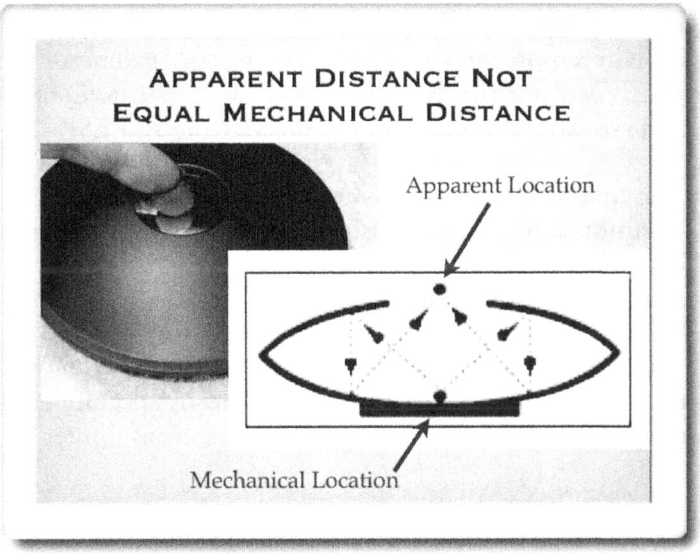

The two bowls are have mirrored surfaces on the inside. The fuzzy ball's actual mechanical location is on the bottom bowl surface. The mirrored surfaces project the apparent image out the center of the top bowl.

When you place your finger on the apparent image, the *apparent distance* between your finger and fuzzy ball is zero. Since the fuzzy ball is actually resting below on the bottom mirror, the *mechanical distance* between your finger and the ball is the distance between the centers of the upper and lower bowls.

In the "spook" illusions and the floating ball illusion, the difference between the *apparent distance* and the *mechanical distance* is created by the geometry of the system.

Eight-dimensional Minkowsky space was selected to explain remote viewing because it creates a geometry which explains the difference between *apparent distance* and *mechanical distance* in remote viewing and ESP.

The first four dimensions are the four dimension of common experience. Classical "lawful" physics pertains to these first four. These dimensions provide the *mechanical distance.*

The next four dimensions of the eight create the *apparent distance.* The "mirrors," metaphorically speaking[33], are the result of having imaginary numbers as dimensions.

Again, metaphorically (not mathematically) speaking, the illusion of *apparent distance* occurs only in certain circumstances. To observe the spooks I had to go into the theme park spook-house system. To observe the floating ball, I had to buy the dish system. In remote viewing and in other examples, the illusion of *apparent distance* only occurs when things are of a *same-system* condition. I will describe the *same-system* condition in a subsequent section.

Also, again metaphorically speaking, since *time* is also a dimension, I can have situations where the apparent spook sitting next to me is from a future or past time. As it was within the theme park ride, it is all a matter of selecting the right set of coordinates, which for remote viewing, are complex coordinates.

Eight-dimensional Minkowski Space

Space-time, Einstein's horizon, is also called "Minkowski space" if, as mentioned before, the time dimension is imaginary. Minkowski didn't stop with four-dimensional spaces, he explored mathematics of spaces with any number of dimensions. There are thousands of peer-reviewed articles in Physics that use various forms of Minkowski space. Here we will talk of "complex 8-dimensional Minkowski space." Complex 8-dimensional Minkowski space uses real and imaginary numbers. That should be acceptable to most people.

The first four dimensional part of 8-dimensional Minkowski space is the space of what we consider the reality of everyday experience. In the following diagram, a set of real "*x*," "*y*," and "*z*" dimensions are shown at the

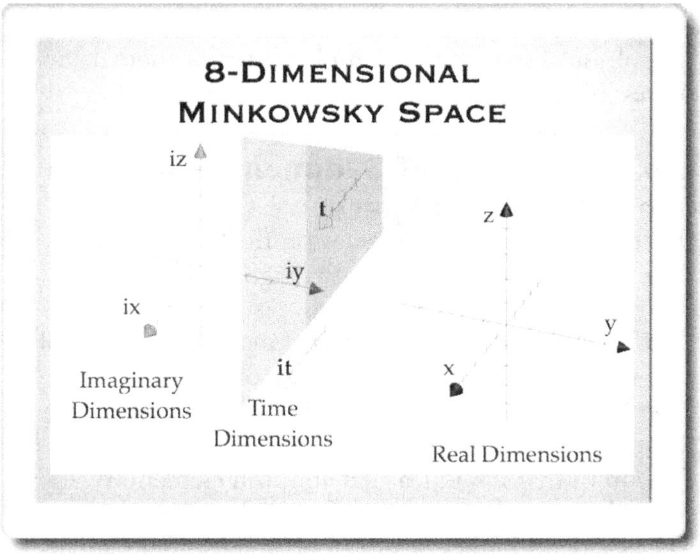

right. In this two-dimensional drawing, I have used up all the three dimensions I can draw in a two-dimensional plane. Mathematically, each added dimension must be at right angles to all of the others. The directions "*x*," "*y*," and "*z*" have used all the coordinates of three-dimensional graphic. For convenience, I represent the fourth dimension, time, in a separate place next to the "*x*," "*y*," and "*z*" dimensions.

Eight dimensions come by saying that each of the four dimensions has both a real and an imaginary part. At the left, for graphical convenience, I draw the imaginary dimensions, "*ix*," "*iy*," and "*iz*" as a separate space. I also add imaginary time next to real time.

Mathematicians have no trouble dealing with 8-dimensional spaces. They are simply subscripts on factors in equations. Mathematicians don't dwell on drawing physical spaces: they simply represent them as matrices in equations.

An Explanation of 8-dimensional Complex Minkowski Space.

First, I show a visualizable conventional 3-space and refresh your memory of the Pythagorean Theorem.

In the "graveyard" of science and mathematics one of the very early "tombstones" is that of Pythagorus. He was a Greek philosopher of about 500 BC. Everyone knows his "tombstone" stands for the Pythagorean Theorem that we all learned about in elementary school. Fewer know that he founded a religious movement called Phythagoreanism. The movement apparently continues today in that there are web sites on the subject.

You probably remember the chant, "The square of the hypotenuse of a right triangle is equal to the sum of the squares of the other two sides." That is the Pythagorean Theorem for two dimensions which is illustrated on the left of the following figure:

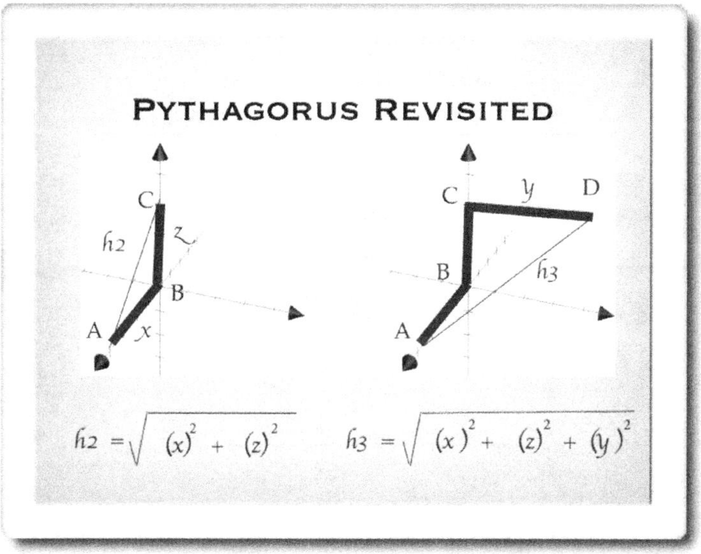

The horizontal axes are "*x*." in this figure. The vehicle axis is the "*z*" axis. In the left hand figure the line from A to B is along the "*x*" axis and has the length x. The line from B to C is up the "*z*" axis and has the length z. As shown by the formula at the bottom of the figure, the distance from A to C is the hypotenuse, h2, and it is calculated as shown by Pythagorus' formula.

Fewer people are familiar with how this works out in three or more dimensions as shown on the right side of the figure. Now we add another dimension. If a line is added going from C to D along the "*y*" axis, a three dimensional hypotenuse line is created. The hypotenuse, h3, the line from D to A, is calculate with the 3-dimensional form of Pythagorus' formula. "The hypotenuse is the square root of the sum of the squares of the sides." This formula still works if we add even more sides.

4-DIMENSIONAL PYTHAGOREAN THEOREM

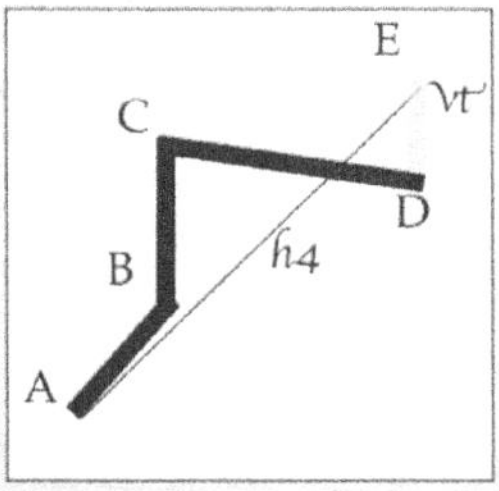

$$h4 = \sqrt{(x)^2 + (z)^2 + (y)^2 + (vt)^2}$$

In this figure, I add another line perpendicular to all the others (which requires a fourth dimension) going from D to E. Here, I have made the line represent time so that I am describing space-time. Since I can't arithmetically add something measured in seconds to something measured in feet, I multiply it by a velocity so that the units work out: velocity in feet per second is multiplied by time in seconds to give an answer in feet. Since my A-B-C-D-E line is representing a four dimensional line, I have to remove the "*x, y, z*" coordinates shown in the preceding diagrams. I can, however, calculate the distance from A to E, the hypotenuse, using the Pythagorean formula with the square of the "*vt*" term included in the formula as shown. I can pick any constant for "*v*" I choose. It is handy to use either "1" or "*c*" (the speed of light), as Einstein preferred.

I could continue this process through all eight dimensions of the complex Minkowsky space.

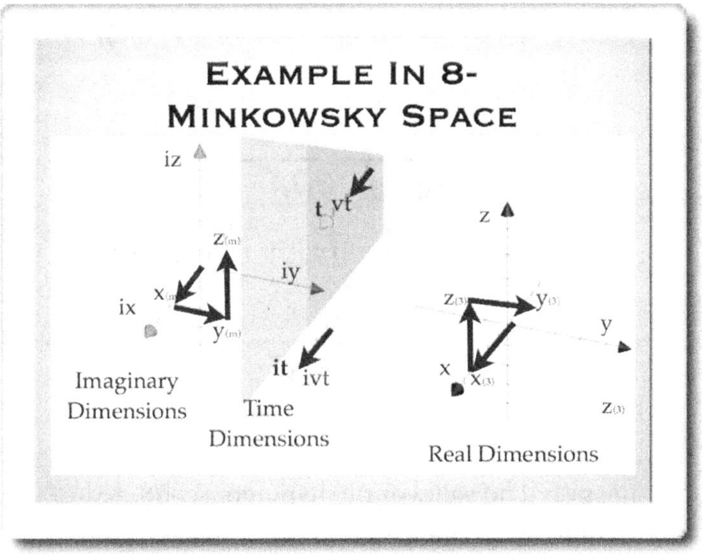

In this figure, I have drawn the components of a complex 8-Minkowsky space line in my two dimensional paper. All these add together in 8-space in a way we can't visualize. I have shown the real components with a subscript (3), for example *x(3)*, and the imaginary components with subscript *m*, i.e. *x(m)*.

The mathematics of 8-dimensional spaces would be very complicated to explain. Fortunately, for this book, I am only concerned with the distance between the start of the 8-dimensional line and its end, the length of the hypotenuse. Again Mr. Pythagorus comes to the rescue. The distance, h(8), is shown in the following figure:

Hypotenuse Length In 8-Minkowsky Space

$$h(8) = \sqrt{((x(3))^2 + (y(3))^2 + (z(3))^2 + (vt(3))^2) + ((ix(m))^2 + (iy(m))^2 + (iz(m))^2 + (ivt(m))^2)}$$

Since $i^2 = -1$, this reduces to

$$h(8) = \sqrt{((x(3))^2 + (y(3))^2 + (z(3))^2 + (vt(3))^2) - ((x(m))^2 + (y(m))^2 + (z(m))^2 + (vt(m))^2)}$$

The formula for the hypotenuse, h8, includes the sum of the squares of all the sides. Since the terms on the second half of the equation contain *i*, the square of *i* becomes -1. The value of the hypotenuse, h8, reduces to the sum of the squares value of the real coordinates minus the sum of the squares of the value of the imaginary coordinates. All the values are conveniently real!

MECHANICAL DISTANCE IN 4-SPACE
MEASURED BY PYTHAGOREAN THEOREM

Mechanical Distance:

$$h(4)=\sqrt{(x(3))^2+(y(3))^2+(z(3))^2+(vt(3))^2}$$

Mechanical Distance in 8-space

In the 8-space, if you set all the imaginary components to zero value, you have the 4-space of ordinary existence. Newton's and Einstein's laws and theories of physics apply. The 4-dimensional version of Maxwell's equations also applies. This 4-space is called a slice through 8-space. Baseballs, rocket ships, flea circuses and cell phones' communications all obey 4-space physics. To avoid confusion, I will refer to 4-dimensional distance as *mechanical distance.* It is measured by the Pythagorean Theorem, in the equation shown previously, if all the imaginary values, e.g. x(m), are set to zero.

The *mechanical distance* in 8-space is the same as the *mechanical distance* in 4-space.

Separating Mechanical Distance From M-Apparent Distance

I will describe the 8-dimensional *apparent distance* as *Minkowsky-apparent distance* or *M-apparent distance* for short.

I will make two postulates: one for Energy, Mass and Time (EMT), and one for *M-apparent (Minkowsky apparent) distance.* I will make the statements only for the macroscopic world to leave out all the mathematical arguments necessary to apply the postulates of quantum mechanics.

Postulate for EMT:

> ***For the macroscopic world,*** Energy, Mass, and Time ***exist only in the 4-space of ordinary experience.***

Postulate for M-apparent distance:

> ***For the macroscopic world,*** information ***can be accessed with zero apparent distance***

M-APPARENT DISTANCE IN 8-SPACE MEASURED BY PYTHAGOREAN THEOREM

M-apparent Distance=

$$h(8)=\sqrt{(\text{Mechanical Distance})^2-((x(m))^2+(y(m))^2+(z(m))^2+(vt(m))^2)}$$

M-apparent Distance

The *M-apparent distance* is shown in the above figure. This is a restatement of the 8-dimensional Pythagorean formula with the real factors summarized by the *mechanical distance* term. The *M-apparent distance* is the *mechanical distance* in 4-dimensions minus the sum of the squares of the values of the imaginary parts.

If any combination of the squares of the values of the imaginary parts adds up to the *mechanical distance* squared, the *M-apparent distance* is zero!

The implications of this are enormous. Suppose an object or person is at the origin of the coordinate system (where all eight coordinate values are zero). They can communicate information with another person anywhere in the 4-dimensional world if their 8-dimensional coordinates are such that the *M-apparent distance* is zero. Communication over zero distance theoretically requires no energy.

In 4-dimensions, communication requires more energy as the separation of sender and receiver increases. The loss of energy goes as the square or cube of the distance, depending on the geometry of the media being communicated through. Everyone knows you have to shout to be heard over distance and, the farther away a person is, the louder one must shout. Radio signals get weaker when we are farther from a transmitting station.

One could ask, "Why can't I buy a cell phone that communicates in 8-space? That way I wouldn't have to keep charging the battery and I could call anyone, anywhere without going through a carrier."

The catch is: at the macroscopic scale, only living things have been demonstrated to use the 8-space medium[34]. Someone with the reductionist scientific viewpoint will ask, "Why is that?" There is no answer for the reductionist. It is an emergent scientifically observed phenomenon, supported by scientific data which agrees with a mathematical model. It is shown in the next section that reductionist explanations are not mandatory for scientific proof.

There is a second catch: living things can use the 8-space medium only when they are part of the same-system.

About Being Members Of The Same-System

> ***Two things can be members of the same-system if they have a zero M-apparent distance.***

Information can be shared by two living things of the same-system if they are electromagnetically coupled or synchronized.

Humans can develop the ability to enter any same-system and gain information at will.

In the case against the validity of remote viewing and ESP, it is cited that no physiological structures exist in the brain which can be identified with remote viewing. That is reductionist thinking. Much of modern physics pertains to emergent phenomena. The issue is of such importance that the Nobel laureate in Physics, Robert Laughlin published a book[35] on the subject.

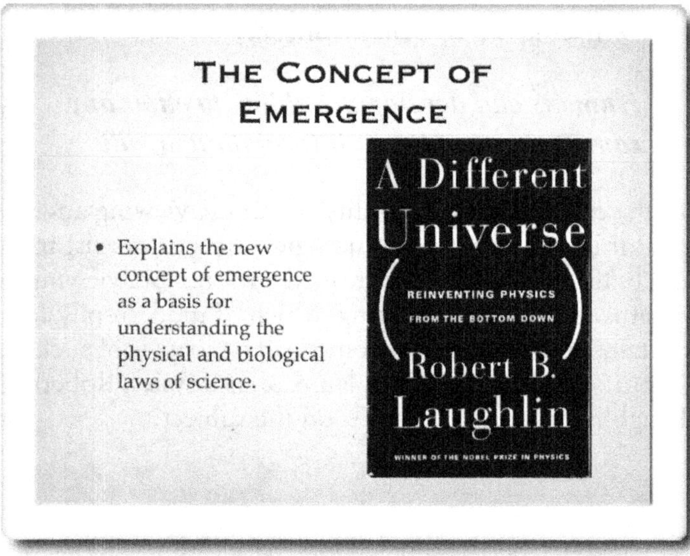

The Concept of Reductionism Vs. Emergence.

Emergent science tries to understand how nature organizes itself and then derive laws of collective behavior. It assumes that physical laws come from collective behavior, not from the behavior of parts.

The concept of reductionism in science is the most familiar in most people's scientific education. The idea began with Newton. Take a problem apart until you have the smallest possible parts: study the parts to find out how they work; study how the parts work together; develop an understanding of how the whole thing works. The reductionist approach assumes that nature obeys physical laws that control the parts and the whole are governed by those laws.

Emergence is a new concept for physics. The above book, *A Different Universe- Reinventing Physics From the Bottom Down*, by winner of the Nobel Prize in Physics, Dr. Robert Laughlin, was published in 2005. In that book he shows us why everything we think about fundamental laws of physics needs to change. We need to learn to accept emergent phenomena and not require a reductionist explanation for everything.

For example, if you want to know why a clock keeps time you take it apart and examine all the gears, bearings, springs and determine their function. Then, you derive how they work together. You identify a part called an escapement, the part that goes "tick-tick," and that part is governed by a pendulum. You know how to calculate how fast a pendulum swings so you can calculate how fast the clock will go "tick-tick." Then, you can calculate the effect of the gears and bearings to find out how fast the hands on the clock move and how the clock keeps time. After that exercise, you can say that the underlaying laws of pendulum motion govern how the clock keeps time.

Reductionistic theoretical physics derives an understanding of the laws of the universe from understanding the laws that govern elementary particles.

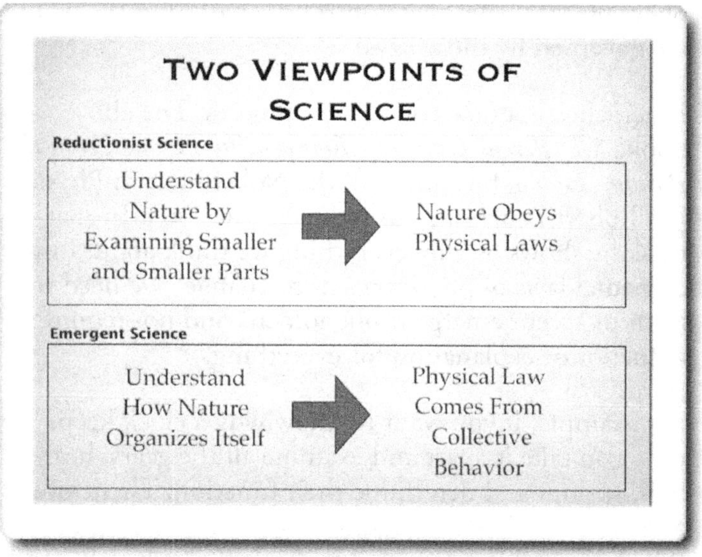

Emergent science, on the other hand, tries to understand how nature organizes itself and then derive laws of collective behavior. It assumes that physical laws come from collective behavior, not from the behavior of parts.

Hydrology is an excellent example of an emergent subject. The system that delivers water to your faucet was designed by hydraulic engineers who know only the emergent characteristics of water. In laboratories, hydrologists have done experiments to calculate parameters that predict the friction of water flowing through pipes and predict the flow of water flowing out of a round hole in a barrel. They deal with behavior of cubic-feet of water, emergent characteristics, not molecules of water. Hydrology deals with the collective behavior of water, not the parts (molecules) of water.

Science does not have a reductionist way to derive the flow characteristics of water by starting from atoms of hydrogen and oxygen, deriving the molecular characteristics and then calculating what happens when it flows in a pipe.

Almost all engineering is based on measurements of emergent phenomena. Hydrology is only one of many examples. The strengths of materials used in tall buildings and bridges are based on emergent measurements. In the satellite business, engineers assemble a spacecraft and then spend nine months testing it in simulated launch and space environments. They do not build a spacecraft, decide it was theoretically sound, and then launch it. They want to know the emergent characteristics of the spacecraft system. There are often big surprises found in testing.

Nearly all, maybe all, of our solid state electronics devices, from your TV set to your PC to your iPod, are using solid state devices which depend on emergent phenomena.

It is important to dwell further on the differences between Reductionism and Emergence because most of the aspects of remote viewing and ESP deal with emergent phenomena. They do not fit into a reductionist paradigm.

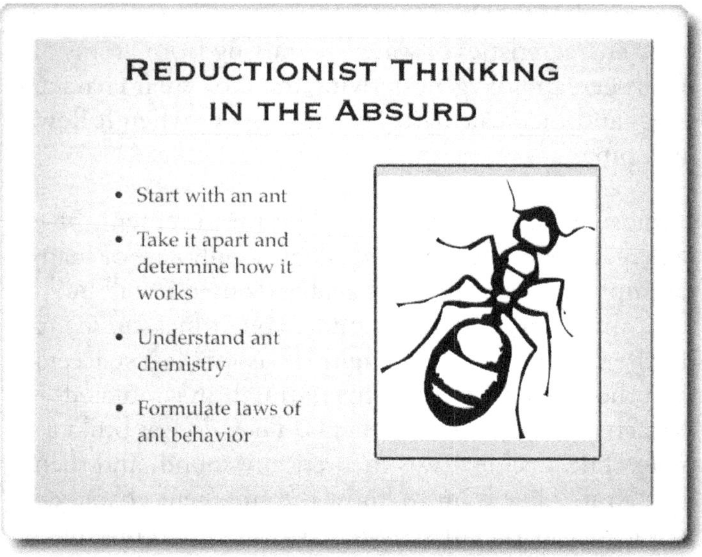

The figure illustrates reductionist thinking if taken to the absurd in the investigation of ants. (Biologists approach their subject from an emergent viewpoint so the example is absurd.) A reductionist approach would be to take the ant apart and determine the functions of its metabolic system and its circulation system; find out what it visual and olfactory capabilities are, and model its biomechanics system. Then, the ability of the ant to do work and it range of travel could be calculated. This knowledge would allow formulation of laws of ant behavior.

This approach is not as absurd as it may seem. I have seen references where investigators have tried to prove the feasibility of bee flight and bat sonar navigation through reductionist techniques.

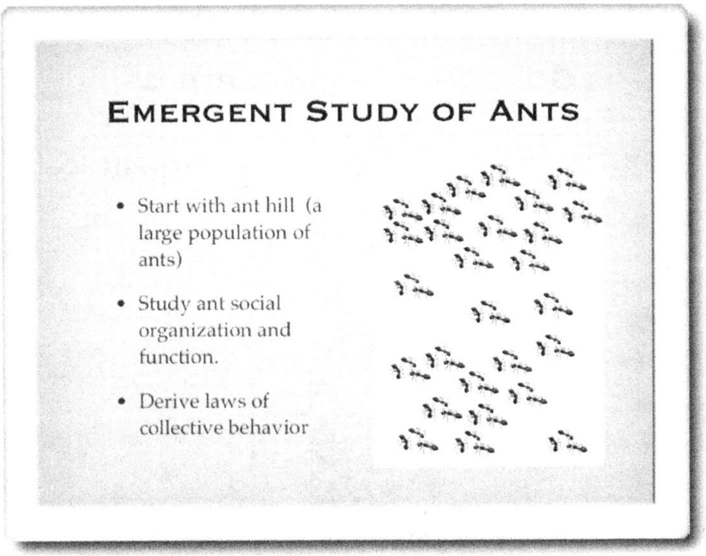

A better study of ants would be the study of ant's group behavior. Start with a large population such as an ant hill. Then determine their collective laws of behavior. The study would answer many questions about ants: the life and death of the colony, the colony as a superorganism, ant communication, and relationships between ants within and between colonies. The study could determine their collective laws of behavior. That study would be the emergent approach.

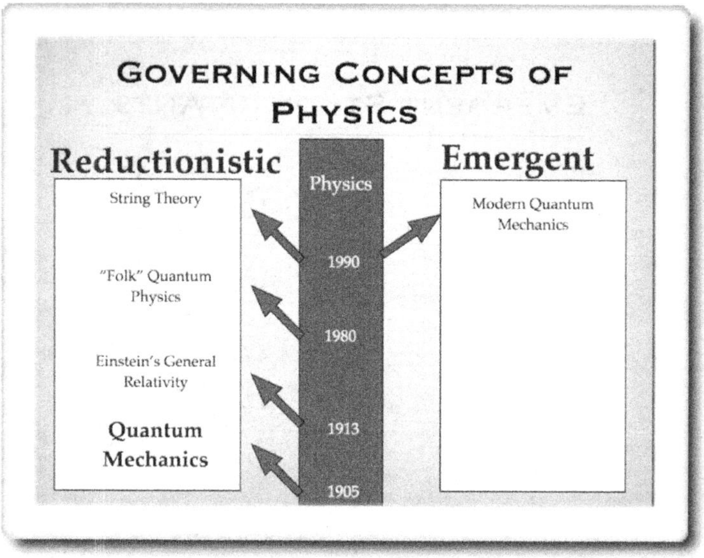

The above tree of selected branches of physics shows which are reductionist and which is emergent.

Early Quantum Mechanics was started by Einstein with his papers on the photoelectric effect where he postulated that light was made up of particles. It became a major subject of physics in the 1920s. The subject concentrated on finding fundamental laws of the behavior of mater smaller than the size of a molecule. Early Quantum Mechanics was reductionistic.

In 1921 Kaluza-Klein theory came into being. Mathematician Theodor Kaluza published a theory that sought to unify theories of gravitation and electromagnetism. He extended Einstein's general relativity to a five-dimensional space-time. With this added dimension, he was able include both the fundamental forces of gravity and electromagnetism. In 1926 Oskar Klein applied Kaluza's theory to quantum theory. The resulting Kaluza-Klein theory paved the way for string theory in searching for fundamental laws governing the universe. This is a classical reductionist path in science and physics.

String theory says that the fundamental particles (quarks etc.) are composed of more fundamental very small strings vibrating in eleven dimensional space. The mathematics of these fundamental parts then are used to describe the whole of the universe from the Big Bang onward. This is the ultimate product of the reductionist approach.

By contrast, current Modern Quantum Physics is in the Emergent column. Most of the recent Nobel Prizes in Physics awarded in this area have concerned emergent phenomena. The physicists' discoveries are of phenomena that are the collective behavior of thousands and thousand of molecules in crystalline structures. The resultant laws of behavior are not applicable to the individual molecules.

For instance, in Robert Laughlin's *Nobel Lecture*[36] he talks about the importance of emergent phenomena:

> "One of my favorite times in the academic year occurs in early spring when I give my class of extremely bright graduate students, who have mastered quantum mechanics but are otherwise unsuspecting and innocent, a take-home exam in which they are asked to deduce superfluidity from first principles. There is no doubt a special place in hell being reserved for me at this very moment for this mean trick, for the task is impossible. Su-

> perfluidity... is an emergent phenomenon... that cannot be deduced from the microscopic equations of motion in a rigorous way and that disappears completely when the system is taken apart... The students feel betrayed and hurt by this experience because they have been trained to think in reductionist terms and thus to believe that everything not amenable to such thinking is unimportant....students who stay in physics long enough to seriously confront the experimental record eventually come to understand that the reductionist idea is wrong a great deal of the time, and perhaps always....I myself have come to suspect that all the important outstanding problems in physics are emergent in nature...."

Clearly, one cannot legitimately argue against the validity of remote viewing and ESP on the basis that it cannot be analyzed by reductionist scientific techniques. Remote viewing and ESP will be shown in the following to be emergent phenomena.

Mechanisms For Biologic Synchronization Provide Physiological Support For Remote Viewing

In the preceding section, it was shown that information can be communicated over large mechanical 4-space distances by making the 8-space M-apparent distance equal to zero. This condition can be maintained when the two entities are members of the same system.

Living Bodies Communicate Information To Other Bodies Electromagnetically Through Cellular Synchronization.

The study of synchronization of mechanically vibrating devices dates back to the Dutch physicist, Christiaan Huygens, who invented the pendulum clock. In 1665 he found that two of his pendulum clocks hanging on the same beam of a wall would become synchronized with their pendulums swinging in unison. Through experiments, he found that a mechanical coupling through the wall beam caused the synchronization. Somehow, the weak vibrations of one clock sped or slowed up the other clock until the pendulums swung unison. Today, we would call this process stabilization by negative feedback.

Incidentally, I have heard sellf-proclaimed "quantum experts" incorrectly make the statement, "quantum mechanics tells us that systems tend to synchronize. If you put two clocks–even digital clocks– in a room and come back later they will be synchronized." They neglect the fact that some sort of energetic coupling, such as mechanical or electromagnetic vibrations, must be present. The clocks must be pendulum clocks mounted on the same beam or digital clocks coupled through an electric field.

Scientists have studied synchronization of mechanical vibrations[37] in cell colonies such as yeast. That synchronization only occurs when there is a mechanical coupling. Mechanical coupling occurs only in 4-space because it involves movement of physical matter according to Newton's laws. Newton's laws have not been shown to extend into 8-space.

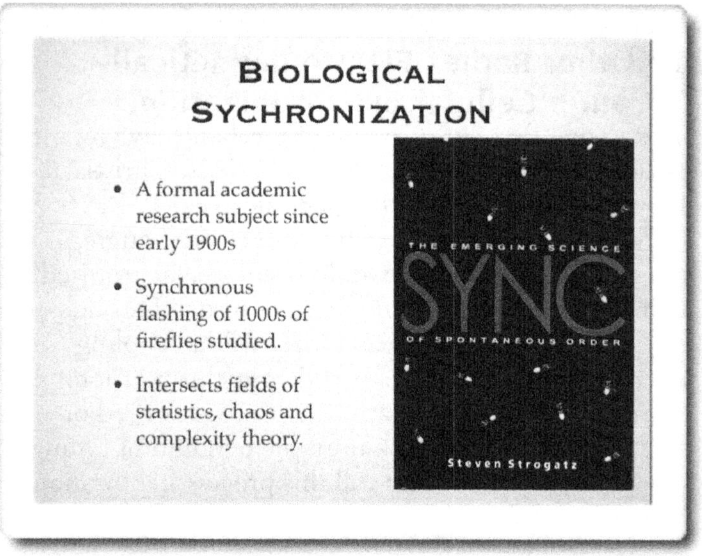

Biological synchronization has become a formal academic research subject since the early 1900s. Mathematician Steven Strogatz explains in his 2003 book,[38] SYNC, that, in the early part of the twentieth century, scientists began to study how whole groups of animals spontaneously organized into doing the same thing together. One such interesting phenomenon was the synchronization of fireflies.

Thousands of fireflies are observed flashing in unison at several places in the world. For example, along a quarter mile of a river in Malaysia or on a whole hillside in Tennessee. They flash like a massive net of Christmas tree lights being turned off and on., There is not a "conductor" firefly orchestrating the flashes. Somehow the whole system of fireflies organizes itself into one time-synchronized unit. First, scientist tried to debunk the phenomena, saying that the synchronization was only an illusion. Then, study showed that it was not one firefly flashing first and then being followed by others, they all flashed together.

In the middle of the twentieth century scientist started explaining the firefly synchronization and other related phenomena. In the 1948, Norbert Wiener published a popular book called Cybernetics or Control and Communication in the Animal and the Machine[39]. It introduced the public to the term, Cybernetics, and the ideas of regulatory feedback in living organisms and machines. The study of Cybernetics led to the fields of Chaos and Complexity Theory. All these areas of research deal with how enormous systems can synchronize themselves. They include dealing with the mechanisms of synchronization of separate biological entities such as people and fireflies and cellular synchronization of brain and heart cells.

Rather than go into all those "tombstones" I can cite core principles.

1. For oscillators, such as nanomotors or nerve cells, there must be:

> 1. A mechanism of communication of the frequency, and
>
> 2. The oscillator must have a way it can be pulled in frequency to allow all the oscillators to self-organize into a coherent group.

Means of communication can be the mechanical vibrations passed through a wooden beam such as described in the previous example of Huygens' pendulum clocks. If one writes the equations of motion for two coupled pendulums, one can find the frequency pulling mechanism.

The communication mechanism in the example of the fireflies is the fireflies seeing each others' flashes. Or, it may be because the fireflies are members of the same system and are electromagnetically coupled with zero M-apparent distance.

In the above mentioned book, *Sync*, Steven Strogatz says there are many conjectures about the frequency pulling mechanism in individuals. The conjectures are hard to prove from a reductionist scientific point of view. The mechanism is an observable emergent phenomenon–that's good enough.

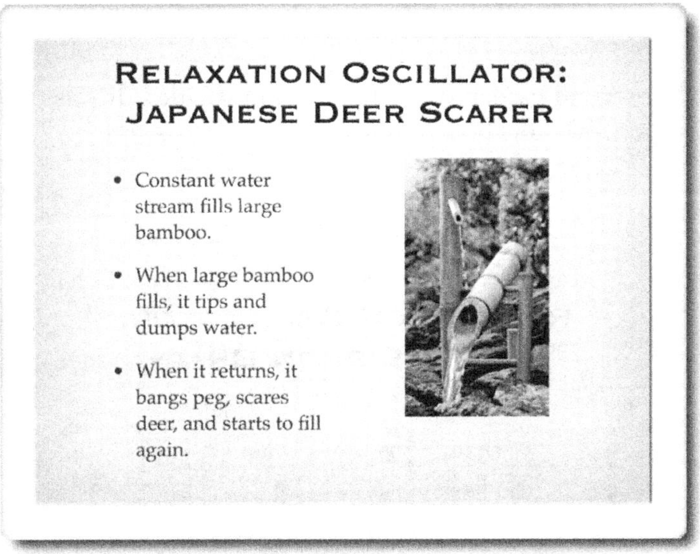

Examples of Relaxation Oscillators

The most common sources of low frequency vibrations in biology are the relaxation oscillators. These are oscillators[40] that mechanically, electrically, or chemically charge for a while and then reach a threshold and discharge with a pulse. Then they start again.

In my garden I have had two good example of how relation oscillators work. The above figure shows a Japanese Deer Scarer, something used for centuries by Japanese farmers to scare deer away from crops.

The oscillation cycle starts with the large section of bamboo in an upright position. Water trickles into the pipe from the water source. When the bamboo gets filled, it suddenly tips over and dumps the water out. Then, it returns to the upright position to start again.

As it does so, the butt of the bamboo bangs against the small peg in the ground making a banging noise to scare the deer.

My second garden oscillator was unintended. It was the water filter for the Koi pond.

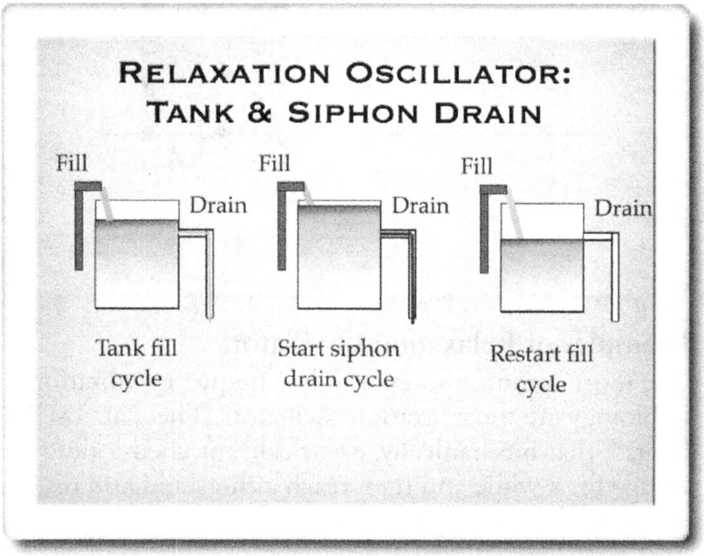

The filter was a hundred gallon tank partially filled with gravel. The gravel was laced with bacteria which digested the fish waste. A pump poured water in the top and filled the tank. A drain took the water to a waterfall into the pond.

Soon after I installed the filter, I found that waterfall would run slow for a two minutes and then there would be a big pulse of water that lasted half a minute. Then it would run slow again. Upon investigation, I found that

I had inadvertently created a relaxation oscillator that ran at a rate of twenty-four cycles per hour. In the above figure, at the left, it shows the tank filling and water slowly going out the drain. As the tank filled, the drain flowed faster. All of sudden, the drain pipe was filled and began to act like a siphon as shown in the second figure. The siphon surge brought the water level down faster than it was being filled until, as shown on the right, the drain sucked air and stopped the siphon. Then, the fill cycle started again.

The Koi didn't like it. I had to study relaxation oscillators to find out how to modify the plumbing.

For the sake of illustration, I could synchronize the filter-oscillator to another oscillator. I could build a big Deer Scarer that dumped five gallons of water twenty-five times an hour. If I arranged it to dump its water into the filter, the filter would reach its level to start siphoning faster and therefor run faster. After a while the filter and the Deer Scarer would be running in synchronism.

Both the Dear Scarer and the filter are nonlinear systems in that they suddenly change modes. The Dear Scarer doesn't move until it suddenly dumps. The filter tank fills gradually and then rapidly drains. Nonlinear systems are analyzed with a branch of mathematics called Nonlinear Dynamics

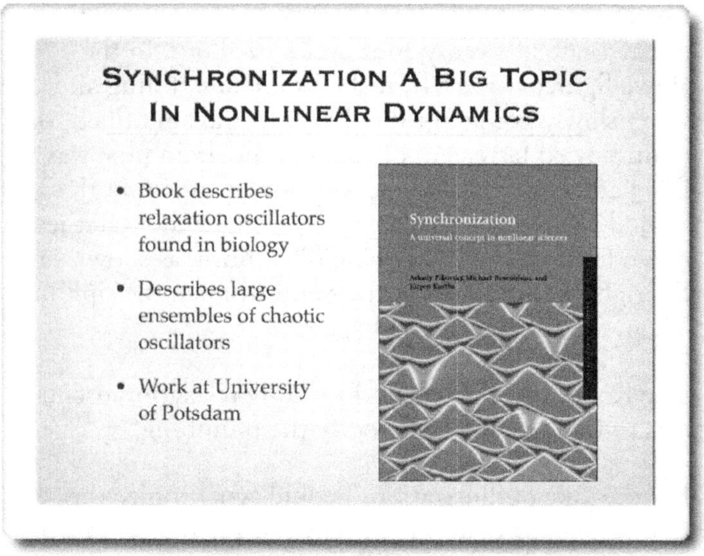

Nonlinear Dynamics of Biologic Systems

Synchronization is a big topic in Nonlinear Dynamics. The above four hundred page book[41] was written by faculty members at the University of Potsdam. It describes synchronization in biology, including singing crickets, cardiac pacemakers, firing neurons and applauding audiences. It describes synchronization of groups of chaotic oscillators and goes into great detail on pacemakers and firing nerves. The first half of this book describes synchronization without formulae, and is based on qualitative intuitive ideas and is readable by anyone somewhat knowledgeable about the technical area. The second half is a definitive book for understanding biologic synchronization for graduate students, and specialist researchers in physics, applied mathematics, engineering and natural sciences.

A shorter paper[42] gives a shorter description on the general mathematical approach. I will not try to give a simplified version of the field because it is too mathematically complex.

For this argument on the validity of remote viewing, it is sufficient to say that biologic synchronization is a well documented phenomenon. There is much published reductionist mathematical analysis of the subject. In a succeeding section, I will show experimental emergent evidence of biologic synchronization of cellular colonies explainable in terms of M-apparent distance.

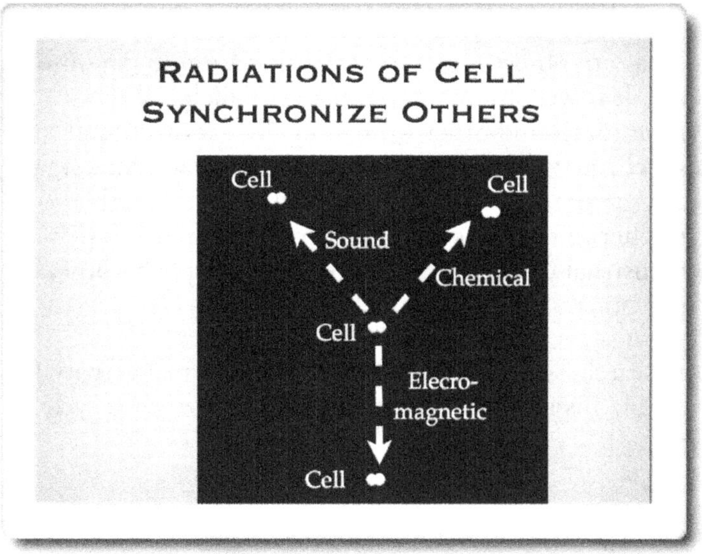

Common Examples of Human Cellular Synchronization

The conventional mechanisms of communication of frequency information for synchronization are mechanical vibrations, electrically conducted signals that pass from cell-to-cell, or radiated magnetic fields from nanomotors or cells like nerves. If thousand of cells are synchronized, conducted signals are measurable by commercially available devices: Electroencephalographs (EEGs), electrocardiograms (ECGs or EKGs), or Electromyographs (EMGs). There are also commercially available devices that sense the magnetic field radiated by synchronized cellular currents. One of these devices is the Magnetoencephalograph (MEG). None of the medical science derived from these measurement systems would exist if there were not demonstrable large scale cellular synchronization.

It easy to observe human synchronization in day-to-day life. It is well known that nurses[43], female college roommates and other women in close contact tend to synchronize their menstrual cycles. I know a kindergarten teacher who said she team-taught with another teacher in the same classroom together all day. Their menstrual cycles were synchronized. People become, in part, members of the same-system.

Researchers[44],[45] have found the brain's alpha activity (8-12 Hz.) is naturally synchronized to the cardiac cycle.

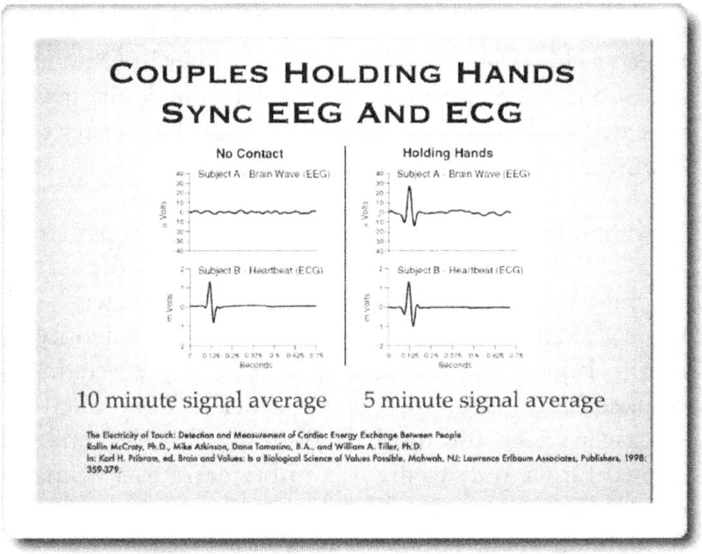

In the studies[46,47] two people sat four feet apart, and EEG and ECG were taken and averaged over ten minutes. The results shown on the left above show that the heartbeat (ECG waveform) of one subject did not appear in the brainwaves (EEG) of the other. After they were holding hands for five minutes a synchronized heartbeat (ECG) signal of one person appeared in the brainwave (EEG) of the other. A synchronization of neural activity was apparent. The two subjects became a same-system by joining hands.

One of the most memorable recorded findings of apparent ESP synchronization is the story of Jim Lewis and Jim Springer[48,49]. Over the years, many people have studied twins separated at birth to investigate the effects of genetics versus environment. The story of Jim Lewis and Jim Springer is an isolated case which is well documented and unexplainable in conventional psychological terms.

The twins, who had been separated by adoption at three weeks of age, rediscovered each other in Ohio at age of 39. Their story was so unusual it made the news and attracted university researchers from University of Minnesota.

Both had been named James by their adoptive parents. In school, both had similar strengths and weaknesses. As children, each had a dog named Toy. Jim Lewis became a security guard, and Jim Springer was a deputy sheriff. Both drove the same type of Chevrolet, chain-smoked Salem cigarettes, drank Miller Lite and liked stock-car racing and football. Both married women named Linda, only to divorce and remarry to a woman named Betty. Both named their sons James Alan, although one spelled it Allan. There were many other similarities in their lives.

Many people drive Chevrolets, chain-smoke Salem cigarettes, drink Miller Lite and like stock-car racing and football, so finding two people with those preferences is not improbable. The other similarities are less probable. There is nothing in the field of genetics that would explain the same names for dogs, children, and wives. For some reason, the two Jims continued in same -system communication after their separation as babies.

Many authors have used the story of the two Jims as proof of ESP. I attribute it to synchronization of thought at zero M-apparent distance. Thousands of sets of twins separated at birth have been studied without finding such a degree of synchronization as the Jims. However, many twins and intimates have experienced otherwise unexplainable synchronization. Most of these data are anecdotal in nature and not what a scientist would call proof.

Many people can give you anecdotal information on same-system ESP synchronizing experiences. Good friends can often finish each other's sentences. I know a couple that often appear wearing the same colored clothes. They say that, although they dress in different closets, their color selection is often the same. Other people will go to work with something important to discuss with their boss, only to find the boss immediately brings up the subject.

I have seen interviews of instrumentalists in great orchestras who work under the tutelage of great conductors. They say that great conductors who are the regular conductors of these orchestras have the ability to communicate their musical intent to the instrumentalists without words or gestures. I recall one violinist, who played under Leonard Bernstein, saying, "If the conductor wanted the violins to cry, they musically cried." This is an example of a conductor creating zero M-apparent distance to his musicians and synchronizing everyone in the orchestral same-system.

Everyone has felt the group energy of a crowd in anticipation of a big concert or sports game. People also feel the group grief of a stunning defeat. Most people refer to this as crowd energy, but it actually is partly synchronization of emotional vibrations. If you have ever performed before a large cheering audience, you know that there is a crowd energy not explainable by conventional physics.

A person that experiences the energies may believe ESP is real. However, it is not a good scientific proof.

This synchronization of people in same-systems theory presents an extraordinary claim. I'll show how it works with well documented scientific experiments.

Body Frequencies,

Until about fifty years ago, all the medical research and diagnosis using EEGs, EKGs, or EMGs was done with signals recorded on paper by pen motors. The mechanical design of these devices, which are like the lie detectors we have all seen in the movies, limited the highest frequency recorded to a few tens of cycles per second. After electronic means of analysis and recording were available, the same general frequency limits were kept to make the data consistent with that taken before. Higher frequencies were considered noise and could be disregarded.

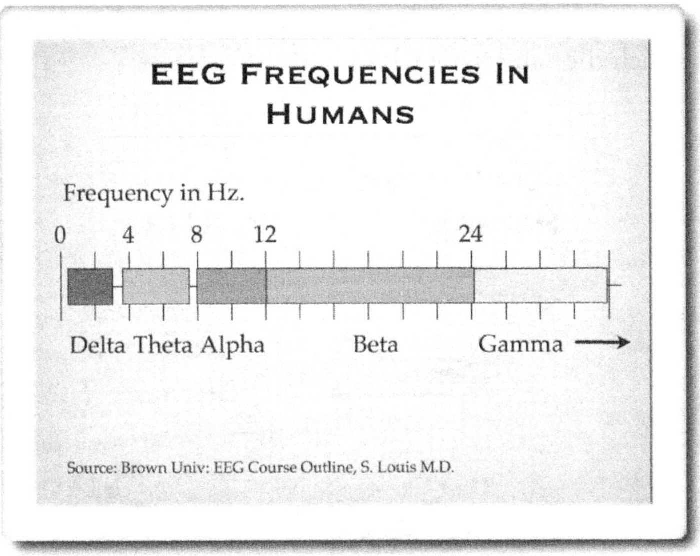

If a signal is to be large enough to be measured by EEG, ECG, or EMG equipment, tens of thousands of cells must be synchronized. Coherence or purity of the signals is attributed to the percentage of cells synchronized.

As previously mentioned, mechanical nanomotors vibrations have been recorded in the 1000 Hz range.

The principle signals of EEGs are in the range of 4 to 100 Hz. The above figure shows the delta, theta, alpha, beta, and gamma frequency ranges. Practiced remote viewers go into a meditative state where the dominant frequencies are in the alpha range. The Monroe Institute, which trained remote viewers[50] for the intelligence program, used a technology called HemiSync. The remote viewers wore earphones with different signals in each ear. The signals were separated by frequencies which produced a binaural beat in the Alpha range to which the subjects synchronized.

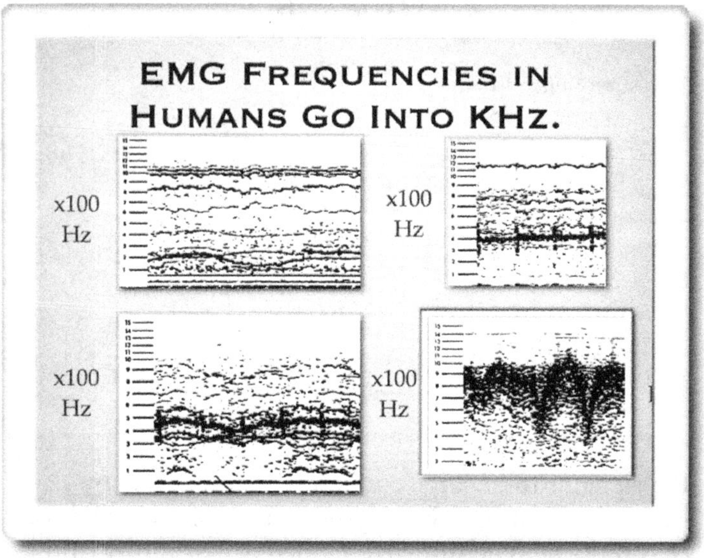

Frequencies That Are Measured By EMGs

In electrical measurement of nerve activity, EMG frequencies go beyond 1000 Hz. In a 1989 book[51], examples of the spectral content of people of different emotional states was shown .

The above figure shows EMG recordings taken from the stomach area. In the left half of the frequency-time figures, the frequency scale goes from 100 Hz to 1.5 kHz in 100 Hz steps. Frequencies are present throughout the whole spectrum. Muscle related frequencies are dominant around the 100 Hz range. The top left figure shows a person at rest and displays many frequencies. There is a slight dominance of frequencies in the 200

Hz range. In the bottom left and top right figures, strong components in the 400Hz to 500 Hz range are observed. In the upper right figure, a strong component at 1.2 kHz is seen The lower right figure shows frequencies spread from 400 Hz to 1 kHz for a person observing images.

These figures and others from the same reference show that there is a great variation of frequencies between people. There are also many frequency components.

Many, if not most, people sense vibrations on others. People will say, "He has bad vibes." I, and many friends who are practiced in sensing the vibrations of others, describe them as in the audio range.

A person with "bad" or angry vibes will convey incoherent, jagged vibrations in the range of below 200 Hz. A happy person has more coherent vibrations in the range of a kilohertz. A very spiritual meditator will have frequencies above 1.3 Kz.

The historical use of the audio band of frequencies is an emergent indicator that these frequencies must be important. Shamen used the low frequencies of drumming and chanting. Gregorian Chants and Buddhist's Mantras are in the low vocal range. Tibetan healers rub healing bowls to make sounds in the mid vocal range. These practices purport to be synchronizing the energies of the body and making the body energies more coherent.

Today there are many recordings sold of synthesized sounds purporting to be the frequencies of the chakras. They are all in the frequency range of the notes on the piano.

A vast amount of work has been done in measuring the electric properties of human tissue.[52,53,54] These are of interest to medical researchers, those who measure electrical signals of fetuses and the human brain. These studies show that frequencies above a few kilohertz are attenuated by body tissues. These data indicate that the low audio signals would have an advantage for electronic intercommunication in a body.

Synchronization Of Cortical Functions

In the above photograph, Joycee and our dog are shown on the coast of where we live. Most people don't realize that the image doesn't exist in the environment. On the left you see a representation[55] of the light wavefront that exists right in front of your nose or in front of your camera lens. The lens of your eye does a Fourier transform of the wavefront outside your eye to form an image on your retina. The lens converts the spatial frequencies of the exterior light into the image. Unless you have a lens there is no image formed.

The image you perceive is not really the image formed on the retina. There is much signal processing in the retina. The optic nerve takes the visual information to the visual cortex and other centers. There it is distributed to many other areas of the brain where specific processing functions take place. One area detects motion, another detects lines, and others detects color, edges, corners, and extract other features. There are thirty specific areas of the brain involved[56,,57] in extracting specific visual information. The image is dissociated into attributes such as lines or colors. These attributes are used by other areas of the brain to recognize such things "this is a face" and then "whose face this is" then "the emotional responses" to this face. Somehow, the information is transformed into the image which I will call "in the mind's eye," your perceived image.

Sensory perceptions such as smell, taste, and sound are also used by remote viewers. These are similarly dissociated and processed in various parts of the brain. Then, like the sense of sight, they are aggregated into nameable entities such as the "taste of a strawberry."

Half a century ago, many researchers were investigating things like spirituality and ESP. One, chemical engineer René Warcollier, did telepathic communication experiments that were somewhat like remote viewing. His experiments had one subject looking at a pencil drawing and another doing what we now call remote viewing, making a pencil drawing of what the other person saw. In his book[58] he shows that in remote viewing, features such as corners, lines or circles can be dissociated from the complete image.

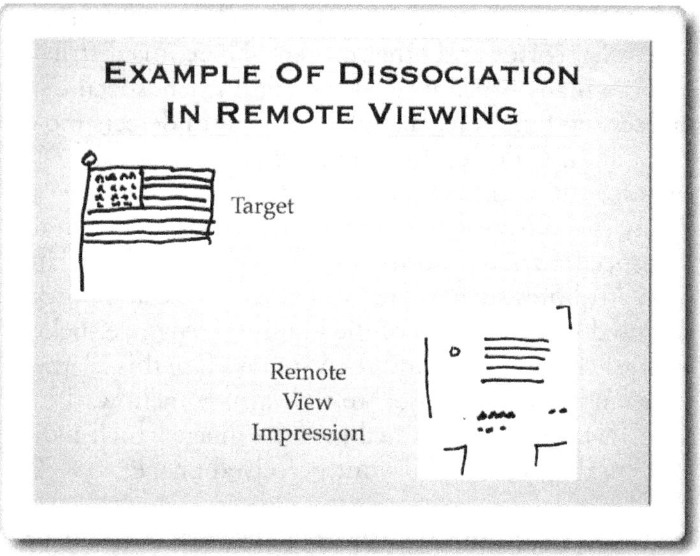

In this figure, similar to one in the book, the target is a rough sketch of an American Flag. The remote viewer drew the elements of the target. However, they were dissociated into features from the original image.

I believe that the remote viewer was somehow able to access the information in parts of the other person's brain where feature extraction is done, but not access the "mind's eye" or perceived image. This is a synchronization process.

Formally trained remote viewers, such as those in the government's intelligence program, start with dissociated impressions of the target and build toward a naming of objects and the environment. They are disciplined to report the sensory data first and later form ideas of objects or environments. They make drawings of dissociated sensed shapes and textures and make notes of sensed smells. They sense activity.

The term "remote viewing" is somewhat inaccurate in describing the activity. They sense all the sensory data at a location, not only the visual information. A better term might be "remote experiencing." A technical description might be "remote sensory synchronization."

For the SRI remote viewing experiments, the remote viewer stayed in an office and an outbounder went to the target site. The remote viewer synchronized his senses (at zero M-apparent distance) with the outbounder in both space and time. They then sensed what the outbounder was sensing and then make drawings and word descriptions of what they sensed. Then, they enlarged their viewpoint in time and space to gain further synchronized data. Remote viewing of this sort is described in *Infinite Mind.*

ZERO M-APPARENT DISTANCE TO FUTURE MEETING

Pythagorean hypotenuse:

$$h(8)=\sqrt{((x(3))^2+(y(3))^2+(z(3))^2+(vt(3))^2)-((x(m))^2+(y(m))^2+(z(m))^2+(vt(m))^2)}$$

View meeting seven days in future:

$$0=\sqrt{((\text{corner of }5\text{th}^2\ \&\ \text{Main}^2)+(\text{in 7 days})^2)-((\text{corner of }5\text{th}^2\ \&\ \text{Main}^2)+(\text{in 7 days})^2)}$$

Synchronization In Space-time

Space and time are a convenient way of not having everything happen at the same place at once.

From an evolutionary standpoint, it is an advantage to have one's perceptions limited to the here and now.

The phenomena of synchronizing in space-time is perhaps the most astounding phenomena of remote viewing and ESP. In the above figure the formula for the M-apparent distance by the Pythagorean Theorem is shown in the upper half. If I want to remote view a meeting I will attend in seven days, I plug in the location (5th & Main) for the *x, y,* and *z* terms and put seven days in the *t* term. Then, my perceptions can synchronize with my perceptions seven days in the future.

One must remember that the information will be vague and dissociated to all except the most talented of viewers. One must go through the remote viewing protocol to access and gain that synchronized information. It is not as though one could send their cell phone camera to the meeting ahead of time. It is more like a series of twitters of dissociated information coming back in time to you. However, by thinking about the future meeting many people might immediately have a general intuition of the meeting. Probably, many people do that quite often, but don't attribute it to remote viewing.

Experiments of viewing pictures in sealed envelopes are examples of synchronizing in space-time. The experiment is set up with a number of distinct target pictures such as the Eiffel Tower or the Lincoln Memorial placed in envelopes. One sealed envelope is randomly selected, and the remote viewer is asked to describe the scene without opening the envelope. The remote viewer records his impression sketches and description in a report. Later, a judge or the remote viewer themselves will compare the report with the picture to determine if the remote viewing was correct.

Research by Courtney Brown[59], described later in this section, has shown that the remote viewer is not viewing the picture in the envelope. They are synchronizing with their selves or the judge at the future time when the report is compared to the picture.

A harder level of remote viewing is sensing a remote site where perhaps only coordinates are provided. It is believed by some, the remote viewer scans and synchronizes with a person who is there, has been there, or will be there in the future.

When I sense the vibrations of someone far away from me, I am synchronizing with their body vibrations described in previous sections. I know that I can do this in real time as well past and future time. For future time sensing I call it intuition on how someone will react to something.

There are many other ESP phenomena which can similarly be attributed to synchronization phenomena.

It must be emphasized that the existence of these synchronization phenomena do not have to be proved at a neurological level to be scientifically valid. That would be a reductionist approach, This synchronization is emergent phenomena.

Effect of Electromagnetic Isolation Chambers On Remote Viewing And ESP.

Many laboratories use electromagnetic isolation chambers in their ESP research and claim that their use proves ESP is not electromagnetic. The person whose psychic or ESP abilities are being tested is placed in a Faraday cage and asked to sense something outside the cage. In the original ESP and remote viewing experiments at SRI, electromagnetic isolation chambers were also used to rule our any sort of fraud[60] where the outbounder and remote viewer might be communicating with radio devices concealed on their body.

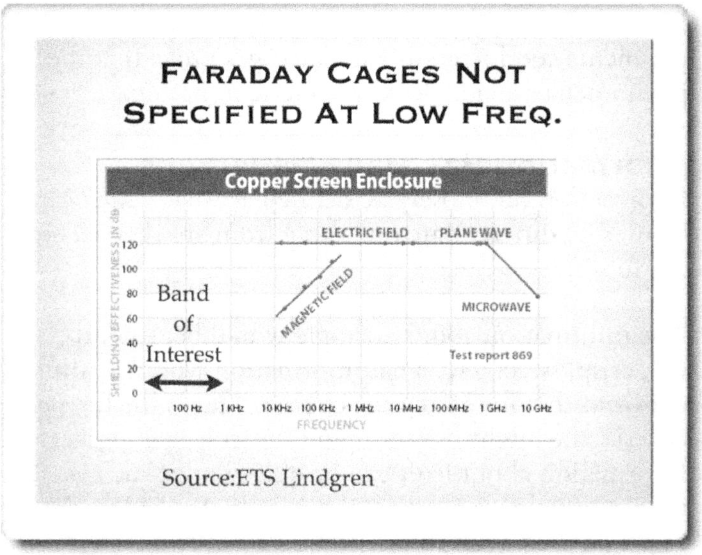

At the frequencies described in the preceding section, a few Hz to a few kHz, Faraday cage isolation chambers are ineffective. For instance, the Faraday cage used[50] in the SRI experiment is ineffective for radio waves of frequency less than 15 kHz, and offers little attenuation of magnetic fields below a few hundred Hz. The above diagram shows the specified frequency response of a chamber in common use.

All the EEG, ECG, and lower EMG frequencies will pass right through the walls of a commercially available Faraday cage isolation room. Shielding of frequencies down to the EEG range requires very expensive, state-of-the art chambers like those that have been constructed for MEG facilities[61]. Those have only been available at very high cost for a few years.

It cannot be concluded that ESP or remote viewing experiments conducted in Faraday cages prove that the phenomena are not electromagnetic in nature.

Information Has Many Definitions

Information can loosely be defined as "that which can be used to differentiate one thing from another according to some criteria."

The information about an apple would be: smooth, red, crisp, sweet, etc. That information would be different from that for an orange or pear. This is the level of criteria one might use shopping for fruit in a market. Information about fruit can be quantitative such as weight or volume if that is the criteria. Sugar content could be used as another criterion.

For biologic cellular communication of vibrations, the basic information is magnitude, frequency, and phase. For two cells to be synchronized they must share the frequency and be phase locked together. In the previously described example of yeast bacteria, cells were locked together in a 1 kHz vibration. The information the yeast bacteria cells shared was frequency and the phase.

For an observer visiting a physical place, such as in remote viewing, the basic information can be sensory data (shapes, colors, sounds, textures, etc.), subjective data (busy, angry, dead), or activities (men chopping trees, automobile traffic.) The information is what makes one place different from another.

In Classical Physics, information can be the state of an object or system. This may be mass, location, spin, velocity etc. This information might differentiate one oxygen atom on one side of the room from one on the other side.

In quantum mechanics, the description goes to a much smaller scale and the information is in the quantum states.

In the field of Information Theory, which is of interest to people who design radio and other communication systems, they define information in terms of signal-to-noise ratio [62,63] and the ambiguity of the information received. For example, there is not as much information in a snowy television picture as a clear one. In a snowy picture, it might be difficult to tell one person from another. The snow is noise and the picture is information. These ideas about information transfer will be important in a later chapter where I discuss accuracy of remote viewing.

Information Is Not Energy or Matter

The important thing to know about "Information" at the macroscopic (not quantum) level is that the description of something is not part of the energy or matter of something. A shopper doesn't physically change an apple by describing it, a lab technician does not change the energy of a yeast bacterium by describing its vibrational frequency, and I don't change the energy of an oxygen molecule by saying where it is. If a remote viewer describes a place, they do not change the place.

Information is not energy. However, transmission of information over distance can use energy. If you use your cell phone to tell someone the color of an apple, energy is required. By conventional physics, and with

Maxwell's Equations particularly, information cannot be transfered over long distances without requiring high transmitter powers. A scientific skeptic would explain that a person in Australia would have to generate an enormous amount of brain power to transmit a vibration that my nervous system could sense in the United States: there is no physical way that could happen. The skeptic would state that ESP sensing of remote vibrations is impossible. They are right, if they constrain their thinking to Einstein's Horizon. However, they are describing the mechanical distance. No power is required to transfer information over zero M-apparent distance.

Extraordinary Evidence For Extraordinary Claims

"Extraordinary claims require extraordinary evidence."

Carl Sagan

The principle of Laplace: "The weight of evidence for an extraordinary claim must be proportioned to its strangeness."

Pierre-Simon Marquis de Laplace

In the previous section, I have presented an extraordinary claim: ESP phenomena and remote viewing have scientific explanations of how they work.

Here I will present extraordinary evidence to support those claims.

Evaluation of Cited Examples

For a theory to meet Carl Sagan's criteria of *Extraordinary Claims requiring Extraordinary Evidence*, it must be at the level of proof of at least *Beyond A Reasonable Doubt.* That is the usual criteria for peer-reviewed publications. It allows future science to advance new theories or discoveries that reduce or eliminate that doubt.

As a practical matter, this peer-reviewed level of proof customarily requires:

1. The claim does not cause the reviewer to loose money[64], organizational support, grants, or product sales if the claim is true.

2. The claim must be within the belief system of the peer-reviewer. People with the greatest disbelief or resistance to something new are those who have to abandon a career devoted to something which the new concept invalidates. This attitude is common in academia and is one of the sources of the Bigot's Protocol.

3. The claim must be supported by an experiment that could be repeated with equivalent results. It would also require that the results would be repeatable by many different researchers.

The remote viewing experiments and operations described in the Introduction certainly fit this level of proof under "3" in the preceding paragraph.

Acceptance of Remote Viewing has been limited by "2." Perhaps the popularization of the 8-Minkowsky scientific paradigm presented by Elizabeth Rauscher and Russell Targ[65] can help to remove these objections.

The examples I cited of teachers who synchronize their menstrual cycles, people who can finish each other's sentences or dress in the same colors, or the amazing story of the two Jim's fit the proof category of *"Probable Cause."* The evidence is circumstantial and anecdotal and might be worthy of further exploration. This is the "where-there-is-smoke-there-might-be-fire" level of proof.

The synchronization of orchestras and "crowd energy" is at the level of *Preponderance Of Evidence"* on my scale, except no one I know of has studied these as a formal ESP phenomena. Conductors and musicians in great orchestras understand and manipulate the synchronization phenomena. Everybody in the business of producing or performing at popular stadium concerts wants to master the manipulation of "crowd energy." Movie directors attend pre-release screenings of their movies specifically to experience the "audience energy" they produce.

If a survey of a large number of people was conducted, there would be evidence that people believed "crowd energy" existed. However, that would be of little interest scientifically without a paradigm that went beyond Einstein's Horizon.

Fortunately, in addition to the remote viewing experience, there are simple experiments that show phenomena that are explained only by 8-Minkowsky space.

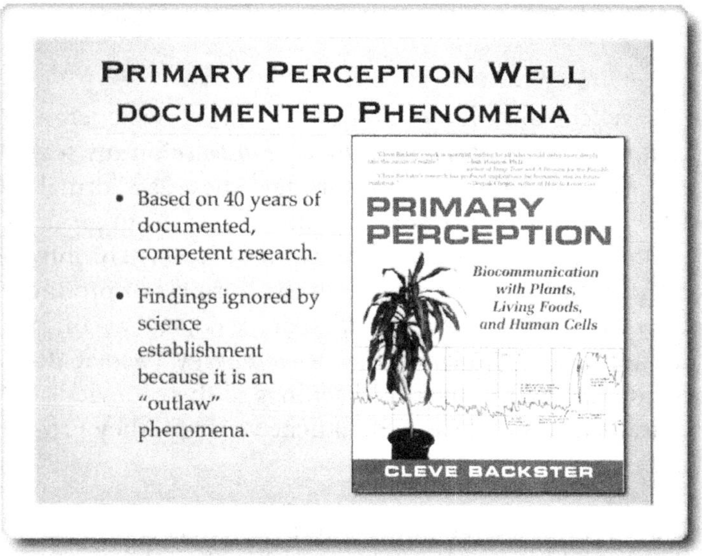

Primary Perception

Cleve Backster published his book Primary Perception[66] in 2003 to summarize his forty year research in biocommunication. His history of scientifically investigating biocommunication and trying to get his work accepted by the scientific establishment is a prime example of how credible "outlaw" research can be unacceptable to those who can't see beyond Einstein's Horizon. Mr. Backster knows the Bigot's Protocol well from receiving it for decades.

If you read about Mr. Backster's background, it is obvious he is not some kind of New Age, unscientific kook or crank. He was one of the pioneers in using the polygraph as a "lie detector." Some called him the Dean of the field. He developed the protocols that are now used in law enforcement interrogations and security checks. He worked for the CIA screening applicants for security clearances for access to the highest level national intelligence secrets. The CIA is very careful about whom they hire. If Mr. Backster was not the very definition of a "straight arrow," conformal thinking, reliable, uncontroversial person he would have not been used to check the reliability of high level government personnel. He was "Dr. Veracity" himself.

After he grew tired of doing routine polygraph examinations, Mr. Backster left the government to form his own business and create more time for research in the field. His private business kept doing polygraphs for the CIA and other government agencies. Eventually, his business expanded into work for commercial business and teaching polygraph techniques. He moved to New York and established a school in Times Square.

His secretary decorated his laboratory with potted plants. In February 1966, he decided to hook up the resistance measuring part of a lie detector to a tall Dracaena rubber plant to see whether he could measure how long it took for water to go from the roots to the leaves. To his surprise, he got polygraph needle fluctuations that were similar to those he observed from humans. He discovered that when he burned a leaf of the plant it responded. Then, he found that the plant would respond to his thinking about burning its leaf. This discovery led to a variety of experiments that showed unexplainable "outlaw" communications between humans and plants. Although he couldn't get the scientific community to have an interest in his "outlaw" activity, the popular press began writing about his experiments. Then, he made appearances on national television shows.

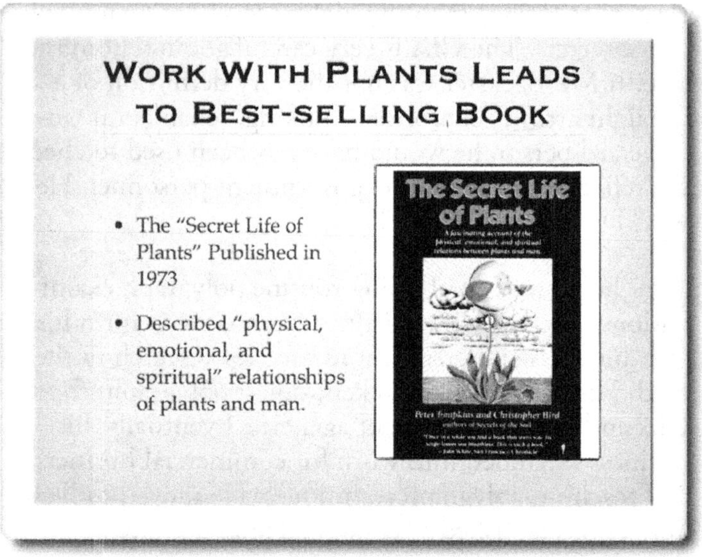

In 1973, other authors published "The Secret Life of Plants[67]" which described his work. It described the "physical, emotional, and spiritual relationships of plants and man" It tended to anthropomorphize plants and attribute them with feelings and intuition. It was a New York Times best-seller and is still in publication today.

That fame made Mr. Backster even more of an "outlaw" in scientific circles. He became "the guy who talked to plants" among parts of the scientific community. However, Mr. Backster was actually quite scientific in his experiments and claims. He decided his plant observations were not ESP since that involves attributing human perceptions to plants which do not have human physiological sensory structures. He invented the term Primary Perception to describe the phenomenon.

He made many additional experiments using brine shrimp, eggs, yogurt, and animal tissues to further explore Primary Perception. Here, I will only use the yogurt experiments as an example. The other experiments are of equal validity. All give extraordinary experimental evidence for exchange of synchronization information between simple undifferentiated organisms over large mechanical distances. This evidence is only explainable by the theory of 8-Minkowsky space.

CHARACTERISTICS OF PRIMARY COMMUNICATION

- Between colonies of undifferentiated cells.
  - No "nervous system" involved.
- Colonies must "tuned"into each other by time spent in proximity.
- Not explainable in 4-dimensional terms
  - No distance squared loss of communication

The Primary Perception Experiments

Backster's primary perception experiments give proof *Beyond a Reasonable Doubt* that colonies of cells can synchronize over distances beyond that which could be predicted by classical physics.

He selected colonies of undifferentiated cells. He did not use animals or plants which had specialized structures such as nervous systems, veins, or leaves. I have chosen to highlight only his yogurt experiments here.

The colonies had to be synchronized to each other by being in close proximity for a while. For instance, Mr. Backster would go to a local market and buy live yogurt. All the yogurt in container would be synchronized to each other.

Mr. Backster would take part of the yogurt from the "home" container and place it in a "roving" cup. He would place two electrodes in the "home" container and connect it to the lie detector. The lie detector would trace a low level background noise. He would then take the "roving" cup and move it across the room, into a different room, or down the hall. He would then poison the "roving" cup with alcohol or some other agent. Soon, the lie detector needle would jump and respond for a couple of seconds indicating electrical activity in the "home" yogurt. The delay of the lie detector response was attributed to the time required for the poison to pervade the "roving" yogurt.

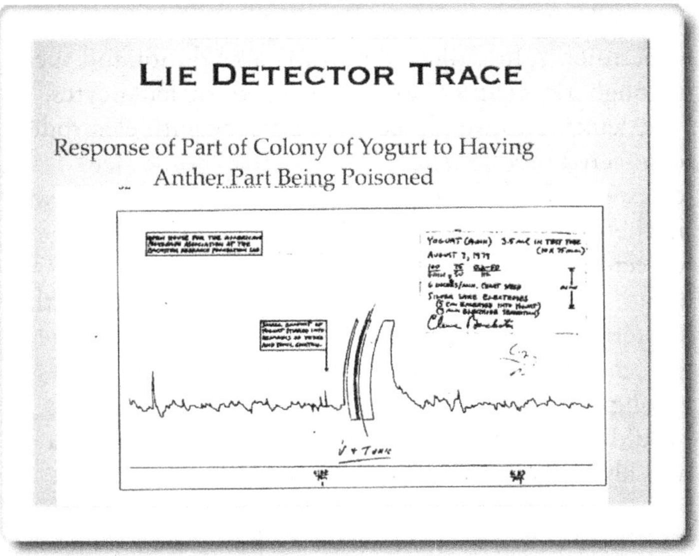

The above trace was taken in during the 1979 American Polygraph Association meeting. An after-meeting cocktail party was being held in Mr. Backster's laboratory. He had set up a demonstration of the yogurt response with the "home" and "roving" yogurt separated by fifteen feet. Everyone in was engaged in social conversation, not paying attention to the yogurt. The yogurt lie detector showed only the background noise response as shown on the left in the above figure. As a joke, Mr. Backster poured his vodka and tonic cocktail into the "roving" cup. About fifteen seconds later, the above response was recorded as the "roving" yogurt died from the vodka and tonic.

Hundreds of similar Primary Perception experiments were run with similar results using a variety of cellular colonies including animal muscle. The CIA was particularly interested in his leukocyte experiment.

Oral leukocytes are white blood cells. In Mr. Backster's experiments, he collected saliva from a donor and then, through a centrifuge process, isolated the leukocytes. The leukocytes we placed in a test tube with electrodes connected to a lie-detector. Here, the “home” leukocytes were in the test tube, and the “roving” sample was in the mouth of the donor. It was found that the lie detector would respond to the emotion of the “roving” donor. For example, when a young male donor saw the centerfold in a Playboy magazine, it registered on the lie detector. Other similar tests found donors, who had strong emotional responses to scenes on a TV set, would produce responses in the lie detector traces. It was also found that the effect was detected when the donor was fifteen miles away from the laboratory where the “home” leukocytes were being monitored. The results of these tests were published[68] in 1985.

Other researchers have replicated Mr. Backster's biocommunication results. Russian experiments confirming biocommunication were described in at 1985 Russian textbook[69] written by a Senior Researcher at the Moscow Academy of Science's Institute of General Genetics. Researchers at the CIA successfully replicated the leukocyte experiments in 1983. It is not known what further research the CIA sponsored. Even as late as 2002, research at the University of Alabama-Birmingham Human Energetics Assessment Laboratory and scientists at the California Institute for HeartMath confirmed Mr. Backster's leukocyte experiments.

Mr. Backster does have a caveat on replicating his experiments. The experiments must be set up so that the effects are produced spontaneously, not at the direction of intention of the experimenter. For instance, reliable results cannot be achieved if an experimenter sits and monitors one culture while another is poisoned. The data are best taken by unobserved recorders and interpreted later. The oral leukocyte experiments work best when the donor has some unexpected emotional surprise. It might not work if the donor were in a laboratory expecting a programmed electric shock.

Apparently, the human observer can "jam" or inhibit the mechanism of biocommunication by closely monitoring an experiment.

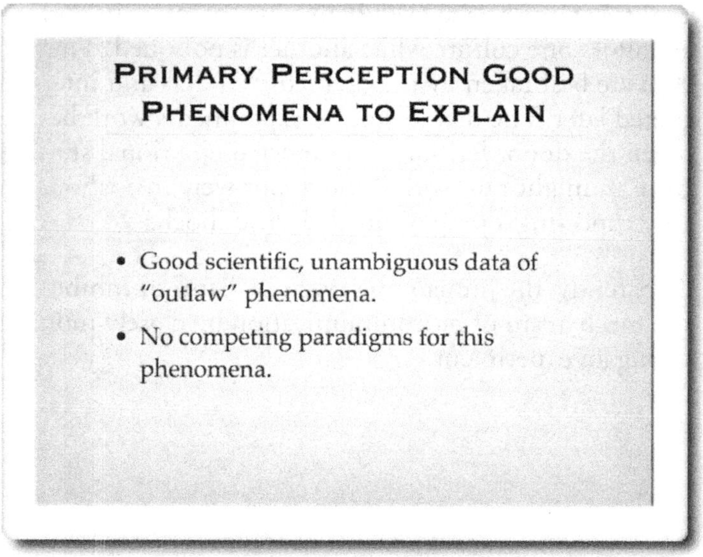

Explaining Primary Perception

Primary perception is a good phenomenon to explain in terms of 8-Minkowsky space and illustrate the idea of zero M-apparent distance. It is also convenient that it a totally "outlaw" phenomenon which has absolutely no explanation in terms of classical physics. There are no competing paradigms to refute or overturn.

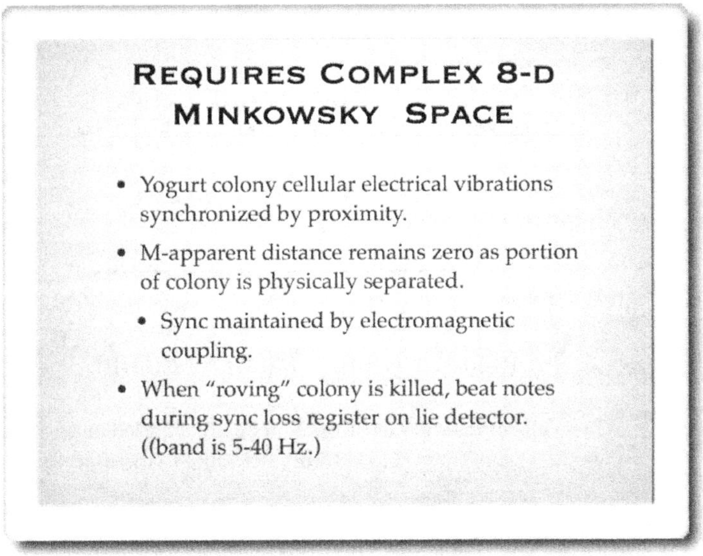

I will start by explaining the yogurt experiment. The experiment requires live yogurt: pasteurized yogurt will not work. The individuals in the live yogurt colony synchronize their electrical vibrations by intercellular conduction and radiation. When the carton of live yogurt comes from the grocery store, the cells will be synchronized. Since all the yogurt cells are electromagnetically coupled, they are thereby part of a same-system. When a portion of the yogurt is spooned out into a "roving" cup, the synchronization is still maintained-it is still part of the same-system

One of the definitions for same-system is that the members maintain a zero M-apparent distance. Recall from the previous discussion that M-apparent distance is as shown in the following figure.

M-APPARENT DISTANCE IS MAINTAINED AT ZERO WHEN CELL COLONIES ARE EM COUPLED

M-apparent Distance=

$$\sqrt{(\text{Mechanical Distance})^2 - ((x(m))^2 + (y(m))^2 + (z(m))^2 + (vt(m))^2)}$$

When cell colonies are electromagnetically coupled and synchronized, the last four terms are equal to the first

When members are part of the same-system, the square of the mechanical distance is equal to the sum of the squares of the last four terms. That equality is what the experimental data shows when synchronization is maintained over great distances.

When the "roving" portion of the yogurt is poisoned, those cells drop out of synchronization and die. It is this desynchronization process that shows up on the lie detector. I will explain the mechanics of that in the next section.

When Backster's work was nationally publicized and popularized, the idea that plants' response could be measured on a lie detector implied that there was some sort of consciousness and emotion involved. That idea really made his work "outlaw." The physiology of plants is well known, and nobody has ever identified a structure such as a nervous system that could support intelligence and emotion.

I theorize that the lie detector response of plants was simply a cellular synchronization phenomenon. The human body has all sorts of measurable electrical activity going on all the time. And, that electrical activity changes with emotion. If a human and a plant are in proximity, some of the electrical cellular activity of the plant will synchronize with some of the frequencies of the human. Some of those frequencies are associated with affinity. When the human drops the affinity frequency, the plant loses synch with them. That loss of synchronization registers on the lie detectors as explained in the following section.

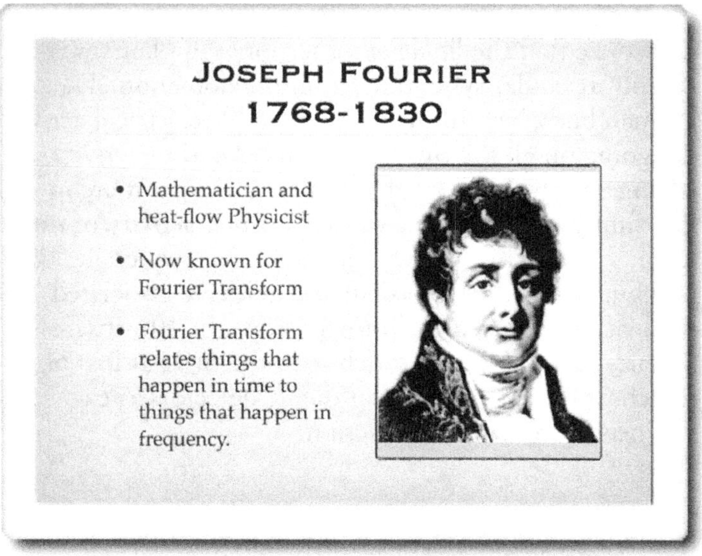

How Biological Oscillators Synchronize

We have to visit another "tombstone" to explain how yogurt biological oscillators synchronize. This is the "tombstone" of Joseph Fourier, a heat flow physicist from the time around eighteen hundred. He invented a technique we now call the Fourier transform to solve heat flow problems[70]. The Fourier transform is a tool to equate an event in time, such as a pulse, to a collection of sine waves which exist over a longer time.

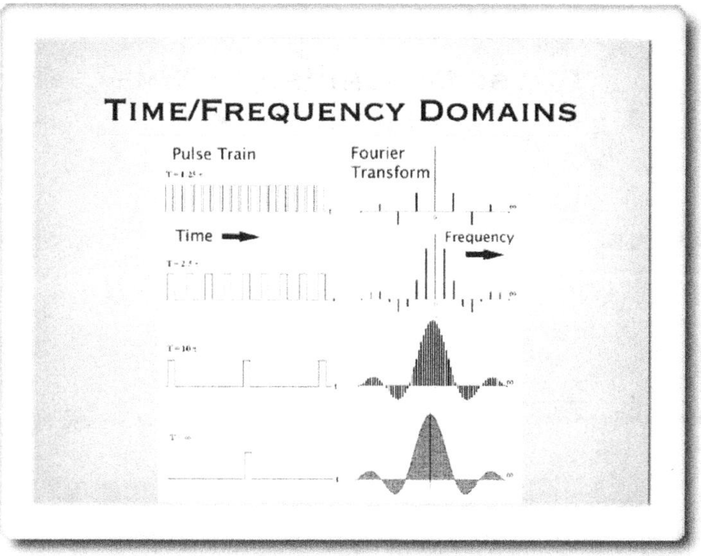

In the above there are examples of Fourier transforms of pulse trains. If the pulses are close together the frequency components are far apart. As the pulses get farther apart, the frequencies get closer together. When there is only an isolated pulse, there is no separation between frequencies.

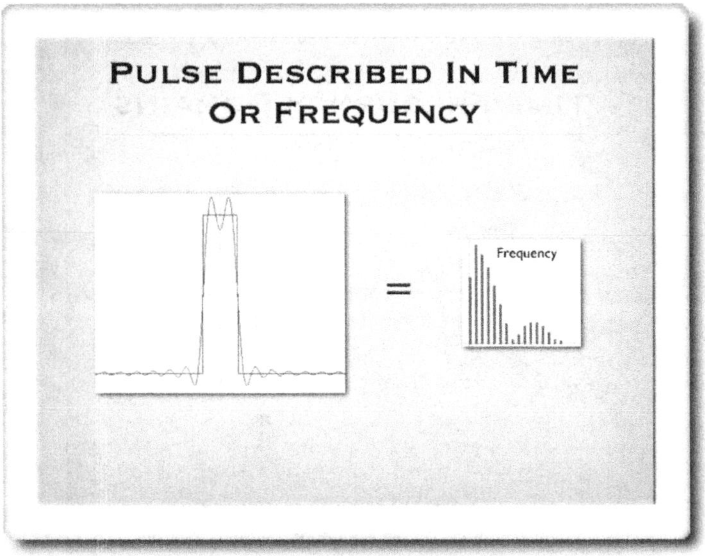

If we look in detail at one pulse in a repetitive series, we see an approximation of the pulse can be made up of fifteen sine waves of the amplitudes shown in the spectrum on the right. The sine waves add up in the vicinity of the pulse and cancel each other everywhere else. The spectrum in the diagram is the Fourier transform of the pulse.

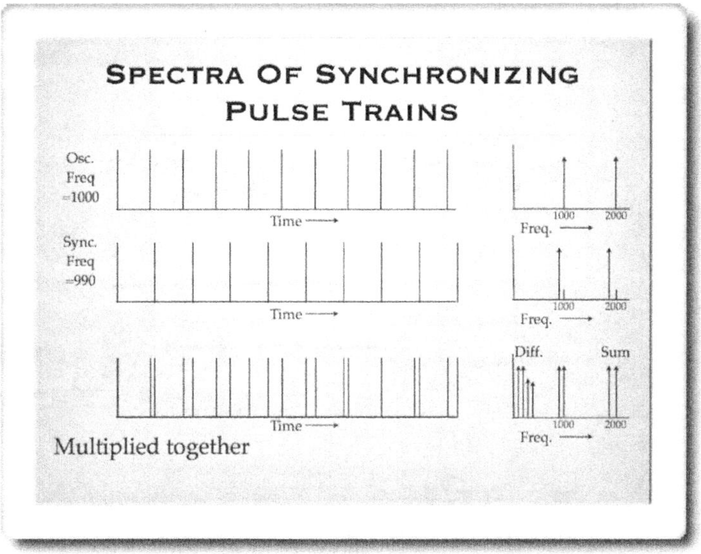

A biological relaxation oscillator, such as described in Section IV, might be operating at a pulse rate of a thousand Hz. as shown in the above figure. The first two of the Fourier transform frequencies are shown at the right at one-thousand Hz. and the first harmonic at two-thousand Hz.

Another oscillator might be producing a series of pulses at a rate of 990 Hz. The frequencies spectrum of that series would start with component of 990 Hz. and have the second harmonic at 1980 Hz. If the two pulse trains are multiplied together, the series and spectrum are as shown in the third line of the diagram. The spectrum of the multiplied pulse train would have the original frequencies plus sums and difference frequencies. In the diagram, I have only shown four of the difference frequencies. The multiplication process produces many more components.

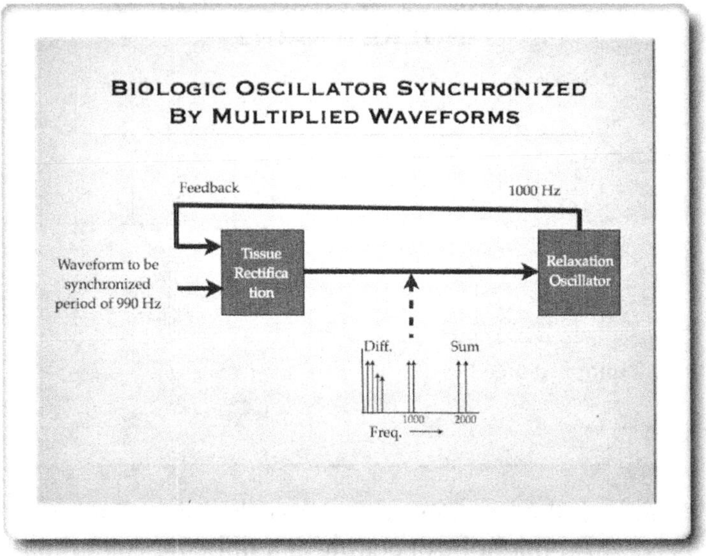

This figure shows how a biologic relaxation oscillator is synchronized. Various parts of the tissue of plants or animals rectify the signals. That distortion produces all sorts of sum and difference components. In the above example, the relaxation oscillator (like my koi filter tank in Section IV) will be pulled in frequency until it is at 990 Hz. Then, the difference components become zero frequency, below the frequency range of lie detectors.

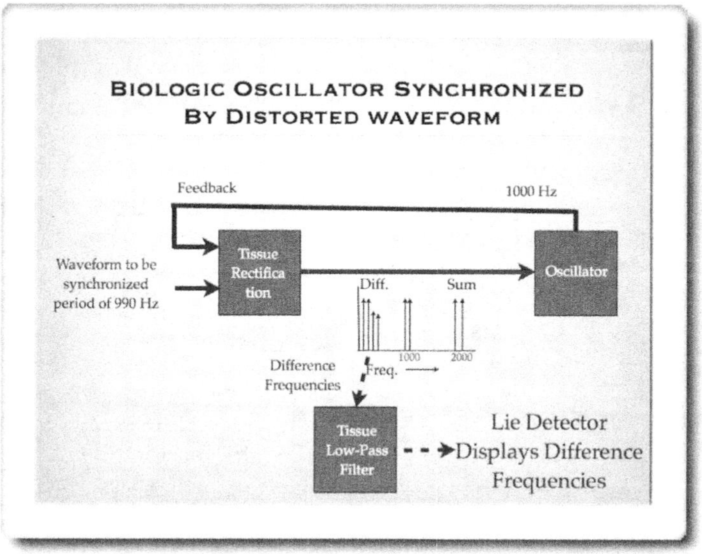

Backster's connected a lie detector to two electrodes immersed in the "home" colony. The detectors use mechanical pen motors to scratch the waveforms on a chart paper. Pen motors do not respond well above 50 Hz. Because the pen motors are electrical, the lie detector input is filtered to the band of 0.2-50 Hz. Since the yogurt oscillating frequencies (and thereby the sum frequencies) are filtered out, all the lie detector responds to is the difference frequency signals.

When the whole colony of yogurt is synchronized (part of a same-system) there will be no difference signals and the lie detector won't wiggle much. When the "roving" part of the colony is poisoned, the dying cells drop out of synchronization as they die and create a difference signal in the "home" colony. The lie detector responds.

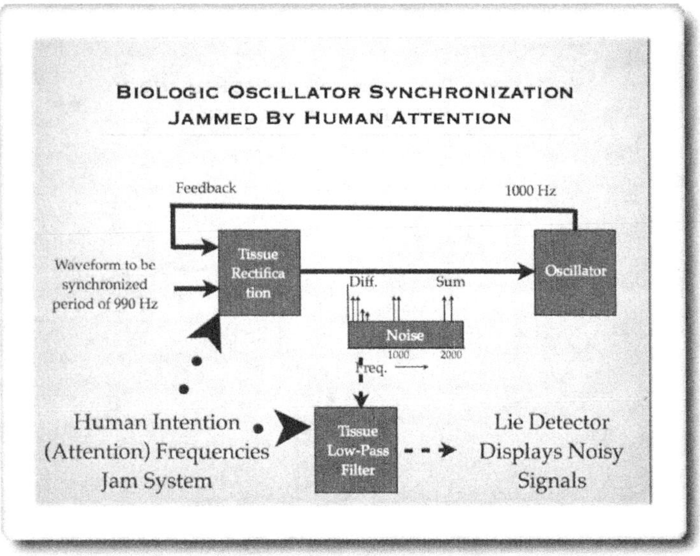

When the experiment of poisoning the "roving" colony is conducted by a lie detector expert who is trained to not put attention on or get involved with his subjects, a good result can be obtained. However, if the experiment of poisoning the "roving" colony is conducted by people who are putting intense attention on the results, or in the presence of camera crews trying to catch the phenomenon on film, the lie detector may not respond. Other scientists may conclude the experiment is not repeatable. There is, however, a good explanation.

When people put attention on the experiment, they make the M-apparent distance from them to the experiment equal to zero. They are unwittingly becoming or intruding into the "same-system" of the yogurt. They add extraneous human frequencies which jam the signal processing. The lie detector will display noisy signals, and the lie detector operator may reduce the noise by adjusting the gain on the lie detector. Then, even when the system is not jammed, nothing will register on the graph. The experiment will not work. This has been demonstrated by Backster and others[71].

The ability for humans to remotely effect cellular functions has been shown by many other researchers. William Braud has shown[72] that humans can remotely change red blood cells.

Proof of Cellular Synchronization

In Section IV, I discussed the concept of Emergence vs. Reductionism. Backster's experiments demonstrated an emergent phenomenon, Primary Perception. I have taken it apart enough to fit it into a general scientific theory. A reductionist would want to "take the ant apart" (the analogy in the section on emergence) to prove what Backster observed is really theoretically possible.

One of the issues is the signal strength, the level of the signals being synchronized, relative to the noise level from the thermal noise and all the noisy stuff that goes on in cells. A skeptic might ask, "how can cells synchronize to the very low signals in the presence of all the noise?" A reductionist would advocate doing a highly mathematical analysis using Chaos Theory and Nonlinear Dynamics. However, from an emergent standpoint all one has to do is put electrodes on a body and observe ECG, EEG, or EMG activity.

Thousands of cells have to be synchronized to produce voltages large enough to be detected. The widespread use of these recording devices is proof of an *Absolutely No Conceivable Doubt* level that electrical cellular synchronization exists.

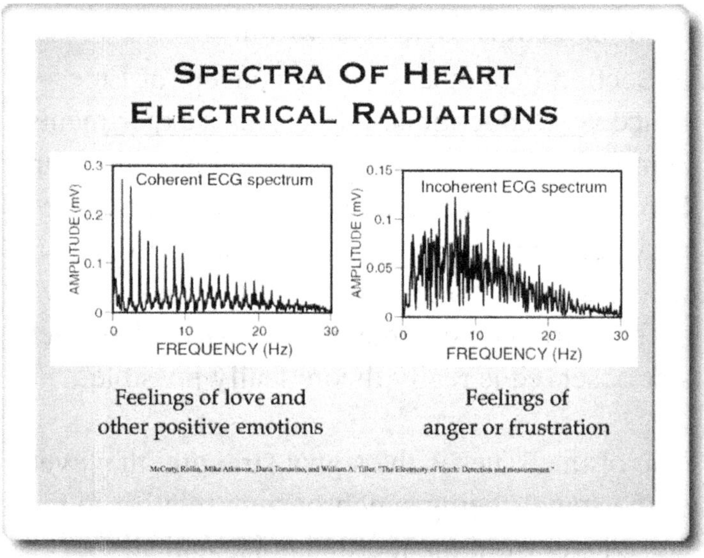

A Hypothetical Mechanism For Plants To Respond To Humans

Studies at the Institute For HeartMath[73] have been performed which show how the spectrum of ECG signal changes with emotion. In the left of the above figure the spectrum of the ECG of a person feeling love and other positive emotions is shown. There are many frequencies to which plant oscillators could synchronize. They could also synchronize to harmonics of these frequencies.

The diagram on the right shows what happens when a person feels angry or frustrated: the heartbeat becomes irregular and the spectrum becomes noise-like.

If a plant were synchronized with these positive emotion frequencies and connected to a lie detector through electrodes, as in Backster's experiments, it would not show significant readings. If something caused the person to become angry, the plant oscillators would lose synchronization and frequency difference signals would be created. The lie detector would show an irregular trace.

The above is a hypothetical example. It is well known from the HeartMath work that people radiate electromagnetic signals as shown above. I know of no experiments synchronizing plants with electromagnetic signals in an instrumented way. This is offered only as a very simplified general indication of how it could work.

From Backster's experiments, it is known that something approximating this happens.

Is Backster's Work Unscientific?

Backster has been criticized by some for his scientific method and lack of publication in peer-reviewed scientific journals. If one examines all his experiments of communication between simple organisms like plants, you do find the use of controls, randomization etc. They are scientific. If one examines only the experiments with plant-human communication, you could conclude they are demonstrations of something, but not scientific tests.

Backster's lack of academic credentials and his primary association with his plant experiments made it impossible for his work to be credible to the Bigots in the "lawful" scientific community. His work was misunderstood and misrepresented in the popular media as proving that plants have or respond to emotions. For instance, on YouTube[74] there is a video of his experiments followed by the narrator (Leonard Nimoy) saying:

> "He also discovered that plants were aware of each other, mourned the death of anything (even the bacteria killed when boiling water is poured down the drain), strongly disliked people who killed plants carelessly or even during scientific research, and fondly remembered and extended their energy out to the people who had grown and tended them, even when their 'friends' were far away in both time and space.
>
> In fact, he found plants can react 'in the moment' to events taking place thousands of miles away. And not only are they psychic, they also are prophetic, anticipating negative and positive events, including weather."

I can understand why university research scientists would not be interested in pursuing scientific investigation of any subject associated with Backster or this description. Researching the subject would invite the Bigot's Protocol and not be a good career choice in academia .

I can see a future university researcher writing papers entitled something like, "Primary Perception As A Demonstration Of 8-Minkowsky Space Synchronization Of Undifferentiated Organisms."

Is Backster;s Work "Extraordinary Evidence?"

A popular interpretation of the meaning of Backster's work would say plants or yogurt colonies, or other tissue colonies display emotion and have some sort of intelligence. The indication of intelligence is an *Urban Myth* level of proof.

I would say it is all a frequency synchronization phenomenon. Backster's work and the work of anyone who records electrical signals from bodies shows *Absolutely No Conceivable Doubt* that cell colonies synchronize.

Backster and other researchers have independently demonstrated that cellular communication can occur at distance. The fact that it can and has occurred is at a level of proof of *Beyond A Reasonable Doubt.*

The hypothesis that Primary Perception experiments do not work in the presence of human attention jamming is anecdotally supported by experience. Experiments need to be done that specifically are aimed at proving that jamming exists. Without such experiments, I would rate jamming as *Probable Cause.*

The Primary Perceptions experiments have always been extraordinary evidence of something. Since there was no scientific explanation of the phenomena it was scientifically ignored. When the ideas of 8-dimensional Minkowsky space and M-apparent distance are added, a scientific explanation exists. Primary Perception becomes admissible as extraordinary evidence.

Overall, I would rate the evidence supporting Primary Perception as *Beyond A Reasonable Doubt.*

Further extraordinary evidence supporting the theory of M-apparent distance is presented in the following section.

The Physics of Non-locality And Distance Invariant Communication

In quantum mechanics, for members of same-systems, everything is happening at the same place and same time.

The Einstein-Podolsky-Rosen Paper

Einstein never bought the ideas of quantum mechanics which were beyond his horizon. In 1935, he took his last shot at disproving quantum mechanics with another one of his 4-dimensional thought experiments. He published a paper[75] with two other mathematicians which showed that quantum mechanics violated one of the conditions of Special Relativity.

In quantum mechanics, there is a concept called entanglement. In a preceding section, I discussed the similar idea of same-systems. Entanglement is the quantum mechanical version of belonging to the same-system. In quantum mechanics two things are the same-system or entangled if they have collided or are from the same-source.

The idea of entanglement is very simple. Suppose that I have a $100 bill and two $50 bills (in my system). I owe $100 to each to Bob and Alice and call them to tell them that I am mailing them either two $50 bills or the $100 bill. When Bob opens his letter and finds the $100 bill he knows (without communicating) that Alice's envelope has two $50 bills in it. The contents of Bob's envelope in one place is dependent on the contents of Alice's at a different place. That dependence represents the idea of non-local entanglement.

Instead of using envelopes containing bills, Einstein's thought experiment used spinning electrons emitted from the same source traveling in opposite directions. This could be like sending one electron to Bob and one to Alice. Because they are from the same-source, their spin characteristics are entangled. That relationship is described in the mathematics of quantum mechanics.

Einstein used the mathematics of quantum mechanics to show that what was measured at one electron (Bob's) effected what would be measured at the other (Alice's). The effect of Bob's measurement had an instantaneous effect on what Alice could measure. Furthermore, if Bob changed the spin of his electron, the spin of Alice's had to immediately change. Since the effect happened instantaneously, the information of Bob's measurement would have to travel to Alice at greater than the speed of light and thereby violate Special Relativity. Einstein felt that this inconsistency shot down quantum mechanics.

Quantum physicists were able to propose some theories that patched up quantum mechanics to explain the inconsistency, Later, in 1964, John Bell restructured the inconsistency into terms known as Bell's Theorem[76] which were amenable to experimentation. Since 1972, many experiments have been run to that proved Einstein wrong and validated quantum mechanics. In a 2008 experiment[77], detectors were separated over 18 km and the experiment was successful in supporting quantum mechanics.

However, quantum mechanics has no explanation for this non-local entanglement other than it simply falls out of the mathematics and can be experimentally demonstrated. Recall that the idea of emergence in physics makes that o.k.

The Rauscher and Amoroso Paper On Relativistic Physics

Recently, Elizabeth Rauscher and Richard Amoroso have provided an explanation for non-local entanglement.

It should be of no surprise to learn that the explanation uses 8-dimensional complex Minkowsky space. The explanation goes beyond Einstein's Horizon. In the terms that I have been using, the electrons in Einstein's thought experiment are part of the same-system and they are separated by a mechanical distance. However, the M-apparent distance is zero so they can interact without violating conditions of Special Relativity.

In my example of the bills and Bob and Alice, they are members of the same-system. In quantum mechanics, states are limited. In my example, there was only a $100 or two $50 bill states available. Since the M-apparent distance is zero, there is only one mailbox. (There is no information separation.) Bob selects one of the letters and opens it. Instantly, only the other letter is available to Alice. That is the idea of entanglement in 8-dimensional Minkowsky space.

In Rauscher's and Amoroso's own words from their paper[78], under the topic of *The Physics of Non-locality And Distance Invariant Communication* they say:

> "Events that appear remote in four space... are contiguous in the complex eight space. ... In this

model, space-time events can become contiguous in the complex eight space, demonstrating that the remoteness of the observer and observed can become contiguous in the complex eight space in which causality conditions are preserved and the acquisition of apparent remote information is allowed."

Are The Experimental Results From Testing Bell's Theorem "Extraordinary Evidence?"

All the experiments run over the last thirty-some years support the refutation of Einstein's 1935 paper and support the idea of non-local entanglement. That is "Extraordinary Evidence" that non-local entanglement is valid.

On my proof scale I rate this evidence as *Beyond Any Conceivable Doubt.*

That evidence also supports Rauscher's and Amoroso's formulation of physics in terms of complex eight Minkowsky space. That *"Extraordinary Evidence"* supports their *"Extraordinary Claim."*

On my proof scale I also rate this as *Beyond Any Conceivable Doubt.*

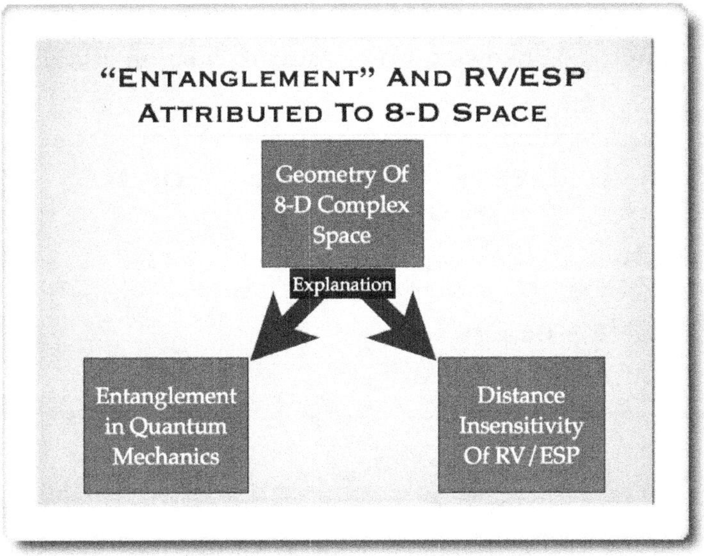

Entanglement and Remote Viewing And ESP Are All Explained By 8-Dimensional Complex Geometry

The geometry of eight-dimensional, complex, Minkowsky space explains explains the distance insensitivity of remote viewing, ESP, and entanglement. Entanglement, and remote viewing and ESP are same-system communications over zero apparent distance that can be attributed to the 8-dimensional complex geometry. That geometry is common to both the microscopic scale of quantum mechanics and the macroscopic scale of remote viewing and ESP.

The experimental evidence showing the distance insensitivity of quantum entanglement, and the experimental evidence demonstrating remote viewing and ESP both show the validity of the assumption that 8-dimensional, complex, Minkowsky space is a representation of reality.

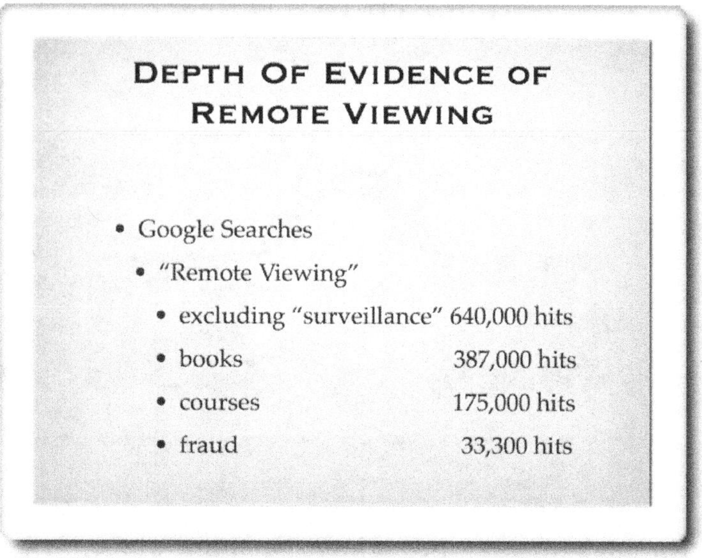

Extraordinary Evidence of Extraordinary Human Abilities

After the remote viewing program was declassified, many former participants joined civilian companies and offered courses for anyone to take to learn remote viewing. Now, on AMAZON there are over 2000 books that are listed after a search of "remote viewing." There are thousands of authors of books about remote viewing and that means there are probably hundreds of thousand of people buying books on the subject.

When I did a Google search of the internet[79] related to "remote viewing", I obtained the above results. I first searched "remote viewing" excluding "surveillance" to eliminate the hits pertaining to video surveillance equipment. The search yielded an enormous 640,000 hits. Obviously, there is much information on the internet about remote viewing. When I searched "remote viewing" and "books," I got 387,000 hits. I was surprised to find "remote viewing plus courses" produced 175,000 hits. "Remote viewing plus fraud" produced 33,000 hits indicating that there are a sizable number of sources criticizing remote viewing.

Amazingly, when I did a search for "remote viewing" services, I got 299,000 hits. There are an incredible number of companies and individuals either offering instruction or remote viewing consultations.

These data are certainly not scientific evidence since the search results include irrelevant subjects. However, it does indicate that remote viewing is not an esoteric subject, believed in by only a few.

By contrast, I searched "wildfire prevention," a very important subject to those of us in California, and got only 147,000 hits. "Death by drowning" only got 136,000 hits. "For motherhood" only gets 196,000 hits.

I can't support "remote viewing" as it is represented on all those sites on the internet. I am sure there many people engaging in spiritual and psychic practices totally unrelated to the SRI experiments and calling it remote viewing. I haven't seen it yet, but I would expect to see someone selling "quantum remote viewing."

The case I will present here only pertains to remote viewing of the type of the SRI experiments, the intelligence programs, and those developed directly from those programs. I limit ESP phenomena to sensing of vibrations, and feelings, and remote medical diagnosis. Those are subcategories of remote viewing as described in *Infinite Mind.*

Validity of Remote Viewing Demonstrated by Length And Funding of Intelligence Programs

Fortunately, the validity and limitations of operational remote viewing have been scientifically established and documented. *Memoirs of a Psychic Spy*[80] describes the operations and, as much as security classification will allow, the intelligence successes from 1979 to 1995. *Limitless Mind*[81] also describes the 1972 to 1995 years of SRI funded research by United States Intelligence Agencies. Government agencies operate such contracts from year-to-year. The agencies, based on results, had to make a decision every year, to continue the programs. That alone, is testimony as to the validity of remote viewing. The length of the programs under continued government support is strong circumstantial evidence that remote viewing works.

The work of William Braud[82] documents scientific experiments that show it is possible to know and influence the thoughts, feelings, mental images of others at remote locations and great distances. The book is based on a collection of papers originally published in scientific journals in the 1980s and 90s. The book describes and evaluates peer-reviewed scientific studies with statistical evidence. They show that remote influences exist and can be validated in experiments.

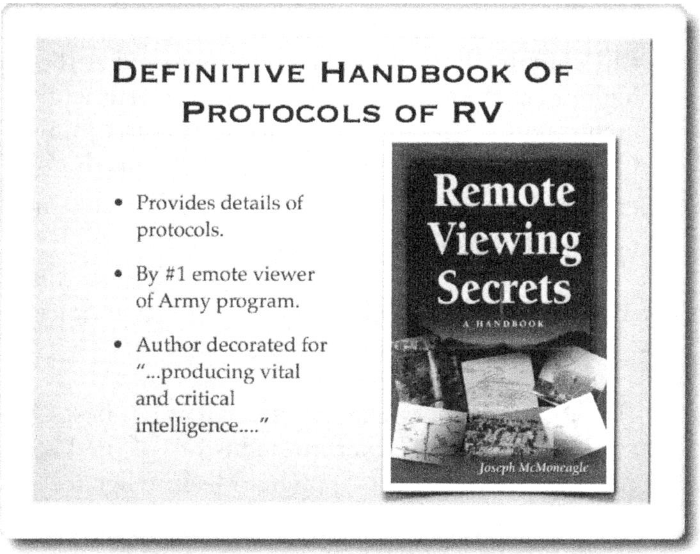

Remote Viewing Secrets: A Handbook[83] was written by Joseph McMoneagle, one of the most effective remote viewers of the intelligence program. In his book, he talks about the program and gives detailed protocols and procedures for remote viewing. Beginning remote viewers must practice intensively for three months to eighteen months before they become an expert. He is recognized for producing detailed remote viewing information. He talks about remote viewing being used to target ideas, concepts, plans, thoughts, feelings, and emotions.

Joseph McMoneagle was awarded the Legion of Merit, America's highest military non-combat medal. His official United States Government citation reproduced in his book says,

> "... he used his talents and expertise in the execution of more than 200 missions, addressing over 150 essential elements of information [EEI]. These EEI contained critical intelligence reported at the highest echelons of our military and government, including such national level agencies as the Joint Chiefs of Staff, DIA, NSA, CIA, DEA, and the Secret Service, producing crucial and vital intelligence unavailable from any other source...."

The back cover of his book *Memoirs of a Psychic Spy*[82] says:

> "Joseph McMoneagle is now known as the best Operational Remote Viewer in the history of the U.S. Army's Special Project–Stargate. His intelligence collection results have never been surpassed and rarely equaled. Among his achievements: He described the interior of a top-secret Soviet manufacturing plant and accurately predicated a new class of submarine under construction there. He sketched the location and described the thoughts of a kidnapped U.S. Army General being held by the

> Red Brigade in Northern Italy. Nearly a year in advance, he accurately predicted when Skylab would leave orbit and where it would impact on the Earth's surface. After conventional reconnaissance failed, he and others were able to locate a downed Soviet bomber that had been carrying nuclear materials. He achieved these results using scientifically designed and tested double-blind protocols."

In his *Remote Viewing Secrets*, he estimates that ninety-nine percent of the material from the intelligence program has never been declassified and made available to the public.

Joseph McMoneagle describes having the Bigot's Protocol run on him throughout his career as a remote viewer. He also details how it was used to end and discredit the program. Highly qualified scientific skeptics with no vision beyond Einstein's horizon and without the proper security clearances were tasked to evaluate the program and prove it was useless. After the program was declassified, their report became public knowledge along with limited remote viewing data. The Bigot's Protocol created the environment for a bunch of career civil servants and military men to use petty fights over turf and budgets to kill the program.

The people who participated in the program and had access to the operational data would rate the veracity of remote viewing as *Beyond Any Conceivable Doubt* on my proof scale from Section II. Many people who did not have access to the operational data would rate remote viewing as *Urban Myth.* Fortunately, recent academic research has provided the openly available data and analysis to support the claims for remote viewing.

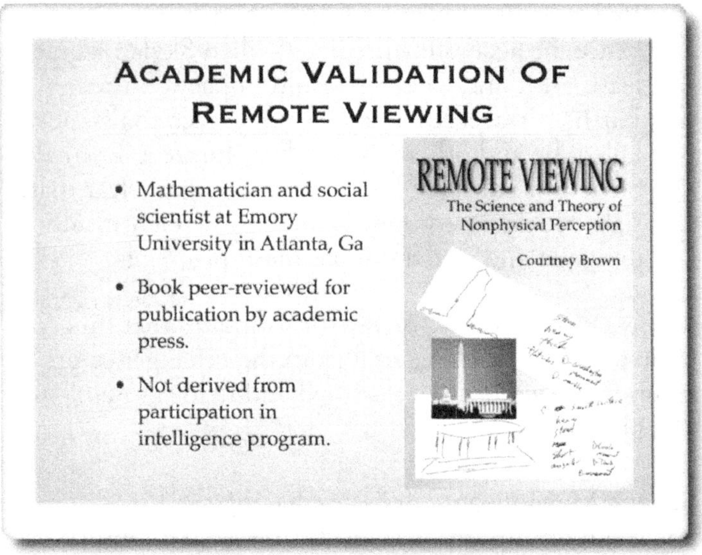

Academic Study of Remote Viewing

Dr. Courtney Brown is a mathematician and social scientist at Emory University in Atlanta Georgia. He has published numerous books [84,85,86] on applied nonlinear mathematical modeling in the social sciences and applied differential equation systems [87]. A recent book [88] is on graph algebra, a new graphical language used for modeling systems.

His book [89] on remote viewing is an equally academic approach to the subject. He developed the book to be published by an academic press. It was peer-reviewed by academics with backgrounds in physics, psychology, social sciences, statistics, and engineering. After a four-year peer-review process and many manuscript revisions, the conservative academic press editors decided the subject was too controversial to be published by them. He decided to publish the book through the Farsight Institute where he is Director.

Although he drew on the work from the intelligence programs, he developed his own procedures for remote viewing. The intelligence programs were designed to produce operational data. They were developed to produce an intelligence product. By contrast, Courtney Brown developed a research protocol with the objective of understanding the *process* of remote viewing. He specifically addressed the issue of selecting or designing good targets for remote viewing. He shows details of his experimental procedures and data and uses a high-powered statistical technique for analyzing experimental results.

To answer the skeptics' usual objections of the experiments being rigged or the data being selected or manipulated after the experiment, he conducted what he called, "a public demonstration of scientific remote viewing." He placed all the experimental data on the internet at every phase of the experiment so that anyone could download it and verify it.

This experiment tested the ability of remote viewers to describe targets that would be selected later in time. The remote viewers viewed a series of fifteen targets and provided their viewing reports in encrypted form. The encrypted data were placed on a website for anyone to download. However, the key for deciphering the data would not be provided until later.

Then, after the remote viewing reports were posted, an independent, unbiased judge, a clinical Associate Professor in Obstetrics and Gynecology at George Washington University School of Medicine. selected the targets for each of fifteen parts of the experiment. That data was published. Then the keys to the encrypted reports were published and the evaluation could begin. The judge made his evaluations and that data was also published. Subjective and objective (e.g statistical) evaluations of the results showed conclusively that in Brown's words, "...the existence of remote viewing is a real, trainable, and increasingly reliable mental phenomena."

In his book, Brown also speculates that higher dimensions may explain remote viewing but, at the time of his writing, he was probably unaware of the work of Rauscher and Targ [90] applying the theory of 8-dimensional Minkowsky complex space. Brown does say that the phenomenon requires more than 4-dimensions. In my terms, he says that we have to go beyond Einstein's horizon. He calls that which is beyond Einstein's horizon "subspace." He goes beyond Einstein's horizon in the path of quantum mechanics. He cites quantum entanglement as evidence of quantum mechanics being able to explain the phenomena. Recall that in this Section, I show how Rauscher and Amoroso can explain entanglement in terms of complex Minkowsky 8-space.

Summary of the Case For The Validity Of Remote Viewing

The opponents stated:

> ***The case against the validity of remote viewing argues that here is no physiological structure to provide mind-to -mind remote viewing or ESP.***

We showed remote viewing is a biologic synchronization phenomena. All the science and medical practices that use ECG, EMG, and EEG devices depend on biologic synchronization. The physiologic structure that provides remote viewing is the brain and the visual cortex specifically. Remote viewing is brain-to-brain communication resulting from the synchronizing of brain functions. Biologic synchronization exists at a proof level of "*Absolutely No Conceivable Doubt.*"

The opponents stated:

> ***The case against the validity of remote viewing argues the claimed distance of mind-to-mind remote viewing and ESP violates the laws of all branches of physics.***

To show that remote viewing has scientific validity we first showed that most people's concept of scientific validity was limited by Einstein's horizon of four dimensional space-time. Extending the mathematics of Einstein's own math teacher, Minkowsky, to eight dimensions allows the apparent distance for brain-to-brain communication to be zero.

The validity of the Minkowsky eight-dimensional mathematics is supported by decades of university physics experiments demonstrating non-local entanglement. In biology, non-local entanglement, described as Primary Perception, has also been experimentally demonstrated for decades.

The applicability of Minkowsky eight-dimensional mathematical concepts to physics and biology has been proven *"Beyond a Reasonable Doubt."*

Many books give interesting examples of somewhat astounding remote viewing successes. These would only be called anecdotal evidence and not scientific proof. The 1976 Puthoff and Targ paper was the first scientific evidence. But, when the remote viewing program morphed into an intelligence program in the early 1970s, it became operationally oriented: viewers gathered intelligence, not scientific experimental data.

Academically oriented scientific research has only been done since about the turn of the century. Courtney Brown's book is the first widely distributed description of a scientific study of remote viewing. He created experiments to test remote viewing. He presents experimental data which are available for anyone's examination. He provides a statistical analysis of that data that is irrefutable. He shows the validity of remote viewing is *"Beyond a Reasonable Doubt."*

The opponents stated:

> ***The case against the validity of remote viewing cites that after spending twenty million dollars of research on remote viewing, the government's own evaluation of the program said it was of no intelligence value. The program was terminated.***

The program director, Edwin C. May, published a paper[91] which he said the scientists who did the evaluation in 1994 did not have the security clearances to access the program or talk to the actual remote viewers. They were only fed data that would support canceling the program. This evaluation was structured by people who had political reasons for killing the program.

Since that evaluation, many books mentioned in the preceding sections illuminated the remote viewing projects and the results. Although the former members of the program are bound by security to not reveal who tasked them and their intelligence products, some unclassified government documents reveal them. For instance, in Joe McMoneage's official citation [92] for his Legion of Merit it is stated he provided:

> "... critical intelligence reported at the highest echelons of our military and government, including such national level agencies as the Joint Chiefs of

> Staff, DIA, NSA, CIA, DEA, and the Secret Service...."

That doesn't seem as though it supports the evaluators' claim that "the program had no intelligence value."

In 1980, during the Iranian hostage crisis, President Carter created a major security breach [93] when he was photographed by the press carrying a folder clearly labeled with the super-secret code name of the remote viewing program, *Grill Flame.* If the President, himself, was reading remote viewing intelligence reports, they couldn't be of "no intelligence value."

Joe McMoneage didn't get the Legion of Merit for doing something of "no intelligence value." Other recipients of the Legion of Merit include Hyman G. Rickover (the "Father of the Nuclear Navy"), Edward L. Beach, Jr. (commanding officer of USS Triton for the first submerged circumnavigation of the world) and John McCain. Joe McMoneage's Legion of Merit peer-group indicates the importance of his work.

One has to conclude that although the Government killed the program for political reasons, it was of great intelligence value at some times. The validity of the remote viewing program is established by The Legion of Merit citation and the photograph of the President. The proof level is "*Beyond a Reasonable Doubt.*"

Conclusion of The Mock Trial Of The Validity of Remote Viewing

I have presented the arguments for and against the validity of remote viewing. I hope that you are now convinced that remote viewing is real and, within the limitations I explained, something that works.

The preceding explanations are necessarily very technical. In a succeeding section, I will present a more simplified view of how it works using the metaphor of an internet browser. This is a "folk" explanation with few footnote references and scientific explanations. I hope this will give you further insight.

---

[19] Maurer, Leon H. "Einstein And The Secret Doctrine." *ABC Images*, Oct 27, 1997.Web. 13 Apr 2009 <http://leonmaurer.info/ABCimages/einstein.html>.

[20] Blavatsky, H. P. *The Secret Doctrine: The Synthesis of Science Religion and Theosophy*, London, 1888. Print

[21] Einstein, Albert. "Zur Elektrodynamik bewegter Körper," *Annalen der Physik* 17 (1905): 891-21.Print

[22] Minkowski, Hermann. "Das Relativitätsprinzip." Undated T. S. . Cod. Ms. Math. Arch. 60: 3. Göttingen:. Niedersächsische Staats und Universitätsbibliothek, (1907) Print

[23] Einstein, Albert, and Marcel Grossmann,. *Entwurf einer verallgemeinerten Relativitätstheorie und einer Theorie der Gravitation* Leipzig: Teubner, 1913. Print

[24] Penrose, Roger. *The Road to Reality: A Complete Guide to the Laws of the Universe.* New York: Knoff, 2005.Print

[25] Rauscher and Targ,"The speed of thought": see citation at # 11 above.

[26] Penrose, Roger. *The Road to Reality*, see citation at # 27 above.

[27] Rauscher, E.A. and R.L. Amoroso, "Relativistic Physics in Complex Minkowski Space, Nonlocality, Either, Model and Quantum Physics" *Physical Interpretations of Relativity Theory*: London, Imperial College, British Society for the Philosophy of Science Conference, 12 Sept. 2008. Print

[28] Rauscher and Targ,"The speed of thought": see citation at # 11 above.

[29] Rauscher, E.A. and R. Targ, "Investigation of a complex space-time metric to describe precognition of the future," *Frontiers of Time: Retrocausation – Experiment and Theory*, ed D.P. Sheehan. Melville: AIP Conf. Proc.. (2006) Print

[30] _______. *Electromagnetic Phenomena In Complex Geometries and Nonlinear Phenomena, NonHertzian Waves and Magnetic Monopoles*, Milbrae:Tesla, 1983, Print

[31] _______. "Non-Abelian gauge groups for real and complex amended Maxwell's equations." Eds. R.L. Amoroso, G. Hunter, M. Kafatos, and J-P Vigier. *Gravitation and Cosmology: From the Hubble Radius to the Planck Scale,* Dordrecht: Kluwer Academic, 2002, 183-188; Print

[32] _______., R.L. Amoroso, "The Physical Implications of Multidimensional Geometries and Measurement", *International J of Computing Anticipatory Systems*, Liege, Belgium. CHAOS,pub. 2006 1, Ed. Daniel M. Dubois, Published by ISSN 1373-541. Print

[33] I am not implying that the dimensions behind the mirror are actually imaginary in a mathematical sense.

[34] With the exception of certain quantum mechanics phenomena where there is a human observer. This is be discussed in Section V.

[35] Laughlin, Robert. *A Different Universe: Reinventing Physics From the Bottom Down.* New York: Basic Books, 2005. Print

[36] Laughlin, Robert. "FRACTIONAL QUANTIZATION." Dep. of Physics, Stanford University Dec. 8 1988 Nobel Lecture, Print

[37] Pelting et. al. "Local Nanomechnical Motion of the Cell Wall Saccharomyces cervisiiae" *Science* 305, (20 Aug. 2004) Print

[38] Strogatz, Steven H. *Sync: The Emerging Science Of Spontaneous Order.* New York: Hyperion, 2004. Print.

[39] Wiener, Norbert , *Cybernetics or Control and Communication in the Animal and the Machine,* Paris: Hermann et Cie - MIT Press,(1948) Print

[40] Kut, Carmen, Vahid Golkhou, and Joel S Bader. "Analytical approximations for the amplitude and period of a relaxation oscillator." *BMC Systems Biology,* 3:6 (2009) doi:10.1186/1752-0509-3-6 (2009). Print

[41] Pikovsky, Arkady, Michael Rosenblum, and Jurgen Kurths. *Synchronization: A Universal Concept in Nonlinear Science.* Cambridge: Cambridge, (2003) Print

[42] Rosenblum, Michael G., Arkady S. Pikovksy, and J Urgen Kurths. "Synchronization Approach To Analysis Of Biological Systems" *Fluctuation and Noise Letters: FLN4.2,Print (2004),* World Scientific Pub.(2004): L53-62. Print.

[43] Klebanoff, Nina A., and Patsy K. Keyser. "Menstrual Synchronization." *Journal of Holistic Nursing,* 14. 2,(1996): 98-114. Print.

[44] McCraty R. and Atkinson M. "Psychophysiological coherence" *Emotional Sovereignty.* Eds. Childre D. and McCraty R, Wilson BC, Amsterdam: Harwood Academic Pub., forthcoming.

[45] ________. "Influence of cardiac afferent input on heartbrain synchronization and cognitive performance" *Int. Journal Psychophysiology*; 45 (2002):72-73, Print

[46] _______. *The Energetic Heart: Bioelectromagnetic Interactions Within And Between People*. Boulder Creek,: Institute for HeartMath, 2003 Print.

[47] _______, Mike Atkinson, Dana Tomasino, and William A. Tiller. "The Electricity of Touch: Detection and Measurement of Cardiac Energy Exchange Between People " *Brain and Values: Is a Biological Science of Values Possible*. Ed.Karl H. Pribram, Mahwah: Lawrence Erlbaum, 1998. 359-79. Print.

[48] Rawson, Rosemary . "Two Ohio Strangers Find They're Twins at 39—and a Dream to Psychologists." *PEOPLE magazine*, May, 1979. Print

[49] Johnson, D.D. "Twins Separated at Birth: The Story of Jim Lewis and Jim Springer." *Smithsonian Magazine* Oct 1980. Print

[50] Atwater, *Captain of My Ship, Master of My Soul*: see #5 above

[51] Hunt, Valerie. *Infinite Mind: Science of the Human Vibrations of Consciousnesses*. Malibu: Malibu Pub, 1989. Print

[52] S. Gabriel,,and. Gabriel, R. Lau, C "The Dielectric Properties of Biological Tissues II: Measurements in the Frequency Range 10 Hz – 20 GHz," *Physics of. Medical Biology*, 41: 2251-69,(1996) Print

[53] G. Martinsen, S. Grimnes, and H. P. Schwan, "Interface Phenomena and Dielectric Properties of Biological Tissue" *Encyclopedia of Surface and Colloid Science*, Marcel Dekker, Inc., (2002) 2643. Print

[54] D, Miklavicˇ,ˇN. Pavs, and F. X. Hart, "Electric Properties of Tissues," *Wiley Encyclopedia of Biomedical Engineering*, Hoboken, Wiley, 2006. Print

[55] The image on the left is the magnitude of the Fourier transform of the photo on the right.

[56] Ramachandran, Vilayanur, Interview "The Human Face", (2001). DVD, ISBN 0-7907-6038-X.

[57] ________ . "40/40 Vision Lecture: Neurology and the Passion for Art." *YouTube.* 04/20/2009). U of California at San Diego. 21 Apr 2009 Web <http://www.youtube.com/watch?v=0NzShMiqKgQ&feature=related>.

[58] Warcollier, Rene. *Mind to Mind.* Charlottesville: Hampton Roads Pub., 1948. Print

[59] Brown, Courtney. *Remote Viewing: The Science and Theory of Nonphysical Perception.* Atlanta: Farsight P., 2006. Print

[60] Puthoff, H.E, and Russell Targ. "Information Transmission under Conditions of Sensory Shielding." *Nature* 251 (1974): 602-07. Print.

[61] Cohen, D, et al. "New Six-Layer Magnetically-Shielded Room for MEG." *Biomagetic Proceedings.*. (Massachusetts General Hospital) 01Jan2002. Print

[62] R. V. L. Hartley . "Transmission of Information". *Bell System Technical Journal.* (July 1928) Print

[63] Shannon, C. E. "A Mathematical Theory of Communication", *Bell System Technical. Journal.*, 27 (1948) 379-423, 623-56,

[64] This is known by some as the First Law Of Science,See Laughlin, *Different Universe,* #38 above

[65] Rauscher and Targ, "Speed of Thought," See #11 above.

[66] Backster, Cleve. *Primary Perception.* Anza: White Rose Millennium P., 2003. Print

[67] Tompkins, Peter, and Christopher, Bird. *The Secret Life of Plants.* New York.: Harper and Row, 1973. Print

[68] Backster, Cleve, and Stephen White. "Biocommunication Capability: Human Donors and in vitro leukocytes." *International Journal of Biosocial Research* 7(1985): 132-46. Print

[69] Dubrov, A.P., and V.N. Pushkin. *Parapsychology and Contemporary Science.* New York and London: Consultants Bureau, 1982. Print

[70] He predicted that greenhouse gasses could cause global warming.

[71] "MythBusters, Episode 31, 6 Sept 2006. Wikipedia. Web. 25 Feb 2009 <http://en.wikipedia.org/wiki/MythBusters_(2006_season)>

[72] Braud, William. *Distant Mental Influence: Its Contributions to Science, Healing, and Human Interactions.* Charlottesville: Hampton Roads Pub., 2003. Print

[73] McCraty, Rollin,et al. In: Pribram, ed. *Brain and Values*: See #50 above.

[74] Backster, Cleve, "Plants Can Sense Human Intention." YEN, YAP WAI . Spiritual Village, 08 May 2008. Video. 25 Feb 2009 <http://spiritualvillage.blogspot.com/2008/05/video-cleve-backster-vs-mythbuster.html>.

[75] Einstein, A., B. Podolsky, and N. Rosen, "Can a quantum mechanical description of physical reality be considered complete?" *Physical Review*, 47, (1935) 777-80. Print

[76] Bell,J.S. "On the Einstein-Poldolsky-Rosen Paradox", *Physics* 1 (1964):195 (1964). Print

[77] Salart, D.; A Baas, J. A. W. van Houwelingen, N. Gisin, and H. Zbinden, "Spacelike Separation in a Bell Test Assuming Gravitationally Induced Collapses." *Physical Review Letter 100*, 220404 (2008). Print

[78] Rauscher and Amoroso. "Relativistic Physics" See #35,above.

[79] Search done March 2009

[80] McMoneagle, *Memoirs.* See #16, above.

[81] Targ, Russell. *Infinite Mind.* See #10, above

[82] Braud, William . *Distant Mental Influence.* See #76, above

[83] McMoneagle, *Memoirs.* See #16, above.

[84] Brown, Courtney. *Serpents in the Sand: Essays on the Nonlinear Nature of Politics and Human Destiny.* Ann Arbor: U. of Michigan P., 1995. Print

[85] _______. *Chaos And Catastrophe Theories.* Quantitative Applications in the Social Sciences series, 107. Thousand Oaks: Sage. 1995. Print

[86] _______. *Ballots of Tumult: A Portrait of Volatility in American Voting.* Ann Arbor: U of Michigan P., 1991. Print

[87] _______. "Differential Equations: A Modeling Approach." Quantitative Applications in the Social Sciences series 150: Sage Pub, Thousand Oaks. (2007) Print

[88] _______. "Graph Algebra: Mathematical Modeling with a Systems Approach" Quantitative Applications in the Social Sciences series 151: Sage Pub, Thousand Oaks. (2008) Print

[89] _______, *Remote Viewing*, See #62, above.

[90] Rauscher, *The Speed of Thought,* See # 11, above.

[91] May, Edwin C. . "The American Institutes for Research Review of the Department of Defense's STAR GATE Program: A Commentary." *The Journal of Parapsychology.* 60(March 1996): 3-23. Print

[92]McMoneagle, Joseph. *Memoirs, See #16, above.*

[93] McMoneagle, Joseph. *Memoirs, See #16, above.*

V Philosophical Trial Scenarios

I will continuing to follow the trial metaphor to discuss two other popular philosophical belief systems that claim to explain why remote viewing and ESP work.

Here, rater than arguing the validity of remote viewing or ESP, the argument is over which is the true explanation of why remote viewing and ESP work. Three sides of the argument share the belief that the phenomenon exist and are valid.

There are many bigoted people who vehemently defend their beliefs that holography or quantum entanglement at a particle level proves the "oneness of all." If you confront them, you get their form of the bigot's protocol.

The "oneness of all" idea has been around for thousands of years in various philosophies and religions. Recently, many have morphed the idea into saying holography or quantum mechanics has proven the "oneness of all." Although holography and quantum mechanics are scientific subjects, their application to remote viewing and ESP is purely philosophical: there is no direct experimental data making the connection.

You, the reader can be the jury. I have already presented the case that remote viewing and ESP can be explained by the geometry of 8-dimensional complex Minkowsky space (in Section IV.)

Here I will examine the two cases that:

> Remote viewing and ESP can be explained by the holographic nature of the universe or,

> Remote viewing and ESP work because entanglement in quantum mechanics works.

Again, the "Levels of Scientific Proof (Belief)" described in Section II should be used for assessing the validity of the premises.

The Holographic Nature Of The Universe.

In metaphysics, some people say that the universe is holographic and that all information exists everywhere. There is a "oneness of all." Some claim that remote viewing and ESP are methods of tapping into the hologram of everything. That is a metaphysical concept that has absolutely nothing to with holograms. It is totally a myth. It is a misrepresentation of the technical aspects of holograms.

For convenience, mathematicians and physicists write equations where the limits of integration (summation) are plus infinity to minus infinity. That convention is of great utility in simplifying otherwise complicated equations. For instance, a term that is something divided by infinity is equal to zero and therefore can be eliminated from an equation. The mathematics of holography uses such conventions. Some people do not understand that, although the mathematics might be valid from plus infinity to minus infinity, the information in a hologram is only local.

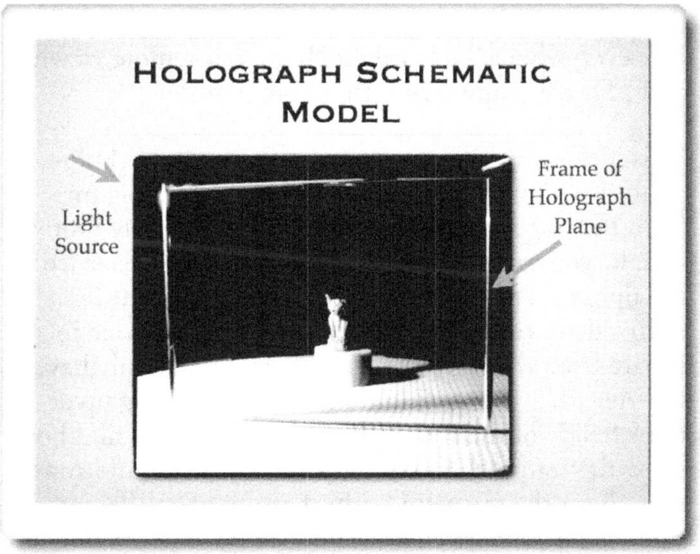

A hologram is essentially a flash photograph of a light wavefront frozen on a piece of film. The above schematic model describes an elementary hologram. There is a small statue on a table. The frame represents the holograph plane where a film would be placed. The exposed film would essentially be the frozen wavefront reflected from the statue. When the film is properly illuminated, a person can see the statue as though looking at it through the same window.

One amazing thing about holograms is that you can cut any one-inch square portion of the film, and if properly illuminated, see the same statue.

Some metaphysicians and "folk" quantum mechanics people say that the picture of the statue is everywhere in every part of the film, therefore, the same information of the statue is everywhere in the hologram. There is a "oneness of all." They then jump to the conclusion that the universe is holographic and all information is available everywhere. They might claim that remote viewing and ESP are tapping into that hologram.

This idea is not even true for the above shown hologram. If you cut a one inch square of the hologram from the upper right hand corner, and properly illuminate it, you can see only the view you would have from the upper corner of the frame: looking down at the right side of the statue. Similarly, if you cut a one inch square from the lower left corner you could only have the view that was available from there: looking up at the left side of the statue. Those two pieces of film both show the statue. However, it is not the same information about the statue in both: they are two different sides of the same statue.

ALL VIEWS OF A HOLOGRAPH DIFFER

The myth that all parts of the hologram hold the same information is wrong.

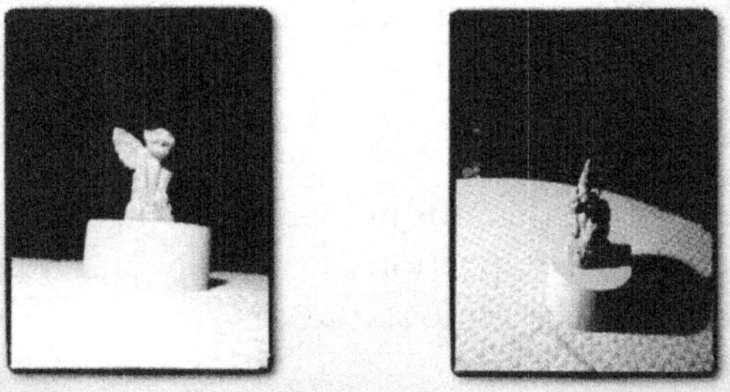

View from lower left corner   View from upper right corner

If the film were twenty feet square instead of as shown, the view from the upper right corner would be of the statue from high above and about ten feet away. That would again be a different set of information.

The wavefront from which the hologram is made must also be described by Maxwell's equations. Maxwell's equations would predict that if the corner on the hologram is farther away from the statue, the statue will appear smaller. To have the same information in the center of the hologram,and far the corner would have to be a violation of Maxwell's equations.

More views of the statue are in the Appendix to further illustrate the concept.

There is no scientific validity to the idea that all information is available everywhere in a hologram. The proof level of the idea that the universe is holographic is "*Not True At All.*"

Because Of Entanglement In Quantum Mechanics

In Section IV, the concept of entanglement and the experimental results from testing Bell's theorem were shown. Entanglement in quantum mechanics at a particle level has been established "*Beyond A Reasonable Doubt*" in many credible experiments.

At a 2008 conference [94], Physicist and author of the best selling book, *The Elegant Universe*, Brian Green gave a keynote talk entitled *The Fabric of The Cosmos: Space, Time and the Texture of Reality*. The subjects he described included string theory and quantum mechanics. He stated and restated unambiguously that quantum mechanics had absolutely nothing to do with anything larger than the microscopic atomic or molecular scale. Many people challenged him. He said some of the metaphysical ideas popularly attributed to quantum theory, may be true if a person has a metaphysical belief system, but, they have nothing to do with quantum mechanics. For example, a metaphysical person may believe they will attract whatever they keep their attention on. Whether or not that is true has absolutely nothing to do with quantum mechanics.

In recent years, a field has originated called quantum biology. It seeks to link quantum physics and the life sciences at a macroscopic level. Many phenomena, such as some forms of photosynthesis, use quantum mechanisms of molecules as part of a larger chemical process. Some others have quantum mechanical conversions of energy into forms that feed chemical reactions. Perhaps one might describe these observations as proof that quantum mechanics is valid at scales larger that molecules. However, there is no experimental link between these cellular level phenomena and ESP.

There are many, today, who say that the proven entanglement phenomena in particle physics is the explanation of remote viewing.

Since such claims are not made by academic physicists teaching quantum mechanics in universities or people writing peer-reviewed papers for journals such as *Physical Review,* I refer to the field of non-technical amateur quantum physics as "Folk" Quantum Mechanics.

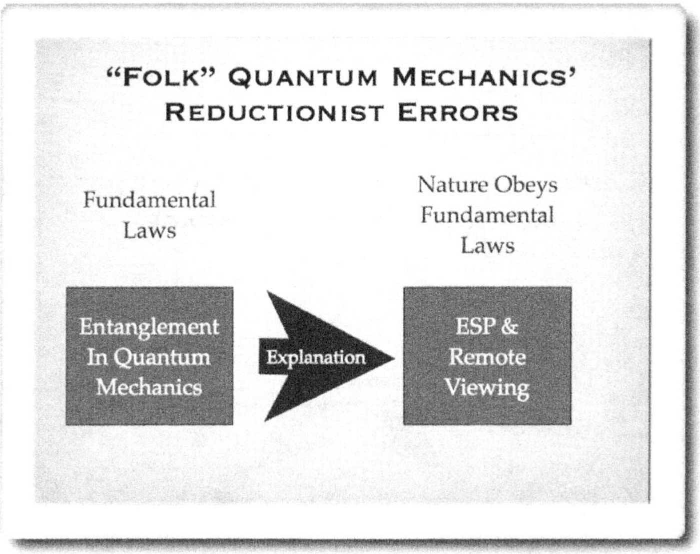

""Folk" quantum mechanics experts will make the reductionist error of explaining that entanglement experiments are fundamental to quantum mechanics. Therefore, they are fundamental to all of nature. Entanglement is experimentally demonstrable in quantum mechanics and, therefore, is the explanation of how remote viewing and ESP can have distance-invariant communications.

That, however, is not correct.

Rather than saying that entanglement explains the distance insensitivity of remote viewing and ESP, one should say that the geometry of eight-dimensional, complex, Minkowsky space explains both phenomena. Entanglement, and remote viewing and ESP are same-system communications over zero apparent distance that can be attributed to the geometry. That geometry is common to both the microscopic scale of quantum mechanics and the macroscopic scale of remote viewing and ESP.

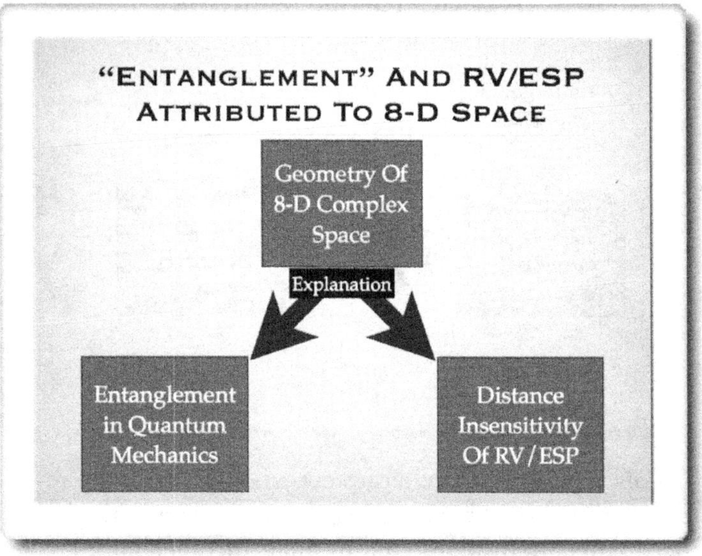

Rather than saying that entanglement explains the distance insensitivity of remote viewing and ESP, one should say that the geometry of eight-dimensional, complex, Minkowsky space explains both phenomena. Entanglement, and remote viewing and ESP are same-system communications over zero apparent distance that can be attributed to the geometry. That geometry is common to both the microscopic scale of quantum mechanics and the macroscopic scale of remote viewing and ESP.

As a bottom line, I have to evaluate the level of proof that entanglement is an explanation for remote viewing and ESP as *True by Analogy* which is slightly above *Urban Myth* on my proof scale.

Some "Folk" quantum mechanics, advocates extends its theories far beyond entanglement. I enlarge on that idea in the following.

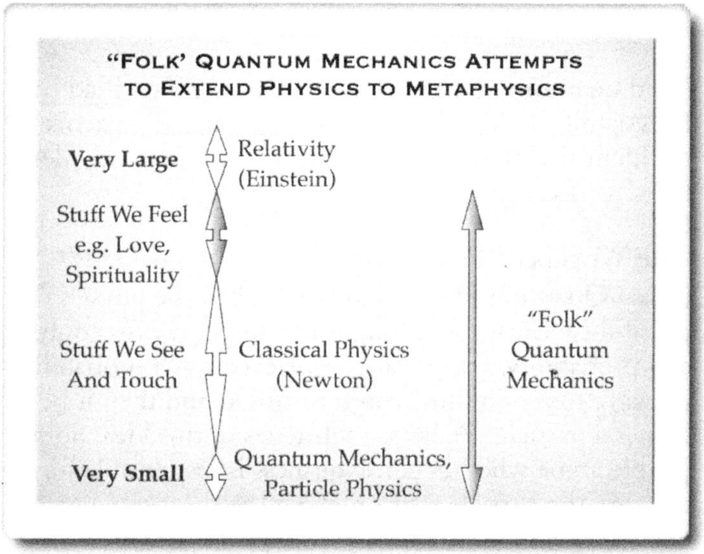

"Folk" Quantum Mechanics

"Folk" quantum mechanics is a viewpoint of quantum mechanics described by people who don't really know anything about it. They have only read about it, or heard about it from other people. Many "folk" quantum scientists, who have PhDs in physics and other unrelated fields, give lectures at metaphysical conferences which bend people's minds with the wonders of quantum physics and how it effects our daily lives.

Today, thousands of products and services are identified with "Quantum" as part of the trademark. A Google search of the word showed 100,000,000 hits. High on the list was "Quantum Performance Tuned Fishing Tackle," and "Quantum Sail Design Group." A book search showed titles such as "Quantum Consciousness: The Guide to Experiencing Quantum Psychology."

The following subjects are common to both "folk" quantum mechanics and academic quantum mechanics.

Valid quantum mechanics statements are identified by (QM) and the "folk" quantum mechanics statements are identified (FQM). My editorial comments are identified as (EC).

The Worlds of Possibility

(QM) Quantum physics has been called the physics of possibility. Until something is measured, it exists only as a probability. For instance an electron is a probabilistic wave function until you measure it, and then it becomes a particle. There are subtleties of this idea. Some people argue whether consciousness is responsible for creating the particle while making the measurement.

(FQM) In our daily lives quantum mechanics tells us that all possibilities exist at every moment and we create the reality by making choices. If I walk down a street and come to a corner, I can choose to turn right or left or go straight ahead. We make thousand of such choices each day and thereby create the reality we experience.

(EC) That is true, but has absolutely nothing to do with quantum mechanics.

Measurement Distorts Reality

(QM) In the famous Heisenberg Uncertainty Principle, a measurement of something like the momentum of an electron can only be done with a mathematically defined uncertainty. The process of measurement alters what is being measured.

(FQM) The "No Child Left Behind" program requires students to be tested to assess school effectiveness. Quantum mechanics tells us that when we measure something we change it. This is demonstrated by the fact that the testing program incentivizes teachers and administrators to drop subjects like art that do not directly contribute to test scores. They change to "teach to the test."

(EC) That is true, but has absolutely nothing to do with quantum mechanics or Heisenberg.

Observation Creates Reality

(QM) Matter doesn't exist until a measurement is made. Before that, it is all a probability

(FQM) Does a tree falling in a forest make any sound unless there is someone there to hear it? The answer is no. If you place a tape recorder in a forest and a tree falls, there is actually nothing recorded on the tape. When you bring the tape recorder home and then start to play it, suddenly the sound will be there recorded on the tape.

As a matter of fact, trees don't fall in the forest when no one is around. They have a state of standing and a state of lying down. They go from the state of standing to the state of lying down instantly in the same manner an electron changes states in an atom. There would be no sound because the tree didn't go through the process of falling. Only if you were watching would the tree go through all those intermediate falling states and thud to the ground for your observation.

(EC) A famous (non-physicist) speaker give this lecture at a metaphysical conference. Over a thousand paid good money to hear this speaker. Of course, none of the tree stuff is true. The reductionist approach assumes that the fundamental behaviors of particles of matter demonstrate a law that, of course, trees must obey.

Schrodinger's Cat

(QM) Schrodinger intended the thought experiment to point out the absurdity of the idea that all probabilities exist until a measurement is made. He is widely misinterpreted.

(FQM) If you really do the experiment with the cat in the box, the fact of whether the cat is alive or dead is only determined by you creating the result when you look inside.

(FQM) In another version of that idea, you apply for a grant to do something. A month later, you get a letter from the grantor. Until you open the letter, both acceptance and rejection can be inside. Your act of opening the letter creates what the letter says.

(EC) The idea that you create your own reality is very common in metaphysics. There are many ways in which that can be true. A person's beliefs and attitudes influence how a person lives their life and the experiences they create. The point, here, is that indeterminacy in quantum theory has nothing to do with what the letter says inside.

"Many Worlds" Conjecture

(QM) In 1957 a paper was presented[95] which said that at each quantum juncture, the universe splits in two and there are parallel worlds with each outcome. If, when you open the box the cat is dead, there is also a parallel world formed where the cat is alive. The person who opened the box also splits in two: one living in the world with the dead cat and one living in the world of the live cat.

(FQM) Quantum theory tells us that at this very moment your body lives in one reality and that right next door there is a parallel universe where dinosaurs munch on prehistoric plants and each other.

(EC) The "Many Worlds" conjecture is only a mathematical idea and an issue for debating the many paradoxes in quantum mechanics. It is a belief structure underemployed physicists delve into. Most of the demonstrations of the phenomena come from special effects studios in Hollywood.

Another common error of "Folk" quantum mechanics is in passing off philosophic ideas and conjectures as being supported by quantum mechanics.

"There is No Out-there, Out-there

In a popular movie[96] about metaphysics, the producers interviewed many physicists and then edited the interviews to support a metaphysical thesis. In that movie a "folk" quantum physicist said, "There is no out-there, out-there."

(FQM) There is no objective reality, only probability. You create reality by observing it.

(EC) The idea that reality only exists when you observe it a distortion of the idea in quantum mechanics that all probabilities exist until a measurement is made. Again, quantum mechanics is only valid at atomic scales. If you stall your car on railroad tracks and you see this big black thing coming at you, it is not wise to sit there to measure the probability that it is a locomotive.

Debates about objective reality versus subjective reality have been around since the time of Socrates twenty-four hundred years ago. For centuries, Hindu and Buddhist teachers have been saying the world is an illusion. This debate doesn't need quantum mechanics.

The Bottom Line on "Folk" Quantum Mechanics

It is all conjectural! All experimental evidence supporting quantum mechanics deals with molecular sized phenomena. "Folk" quantum mechanics is a branch of eastern philosophy, not of science.

"Folk" quantum mechanics mistakes metaphors of scientific phenomena for science itself. The Uncertainty Principle of quantum mechanics pertains to atoms. Whether or not a falling tree makes noise if there is no one there to observe it has nothing to do with quantum mechanics.

"Folk" quantum mechanics is a confusing attempt to relate metaphysics to a scientific paradigm. "Folk" quantum mechanics has only added an "in" technical jargon to an old metaphysics. If you can say, "I brought the idea up from the quantum foam," or "I released the idea into the quantum field," you can be in the select group considered to be at the leading edge of metaphysics.

"Folk" quantum mechanics has nothing to offer in explaining remote viewing or ESP.

---

[94] ISSSEEM 18th Annual Conference: Energy, Intent, and Healing.Boulder CO, June 2008

[95] Everett, Hugh, III."'Relative state' formulation of quantum mechanics." *Review of Modern Physics* 29:(1957): 454–62. Print.

[96] I can't mention the title for legal reasons.

VI Mind-as-browser Metaphor For Remote Viewing

Here, I will describe some aspects of the function of remote viewing using the metaphor of the internet browser. I do not imply that this is the way the mind really works.: There is absolutely no scientific data that the mind works this way. This metaphor will help you understand the *function*, not the mechanism, of remote viewing. I have created this metaphor to integrate what I learned from reading all the previously cited references.

The internet is a global source of information including things that are occurring in real time and records of things from the past such as documents and YouTube videos etc. To access information you need the URL, the Uniform Resource Locator to put in an internet browser such as Firefox or Safari. The URL contains the protocol (e.g. http//:) and the domain (e.g. whitehouse.gov) which, somewhere on the internet, is converted to a physical server address (IP address), which is kind of like a computers' telephone number. The IP address also directs your inquiry to the physical location on that server's memory where the file is stored.

After your browser connects to the server file, the data are communicated to your computer and displayed. How fast that happens depends on the bandwidth of the server and the internet connection.

Here the metaphor of the browser and internet is used to explain the *function,* not physical reality of remote viewing.

The Extrasensory Resource Locator

When you use your computer and internet browser, you place a URL in the window blank to start acquiring the information of the website.

When the mind acts like a browser, instead of a URL there is an ERL (Extrasensory Resource Locator). It can be:

Space-Time Domain (physical locations in *x,y,z*, and *t*)

> ***Place*** (in current time) such as Detroit or Longitude 120 degrees west, 34 degrees north. Or it can be what is in my pocket or in a paper sack.
>
> ***Time*** such as last Tuesday at noon, or next Tuesday at the close of the Stock Market.

Memory Domain (subjective locations and times)

> ***Events*** such as what you were doing when you saw the TV coverage of 9/11. Or, it could be when you fell skiing and broke your leg or when you crashed your motorcycle.
>
> ***Emotion*** such as the last time you felt immobilizing fear.
>
> ***Sensory*** *event* such as the last time you smelled a skunk or a dead animal.

The Space-Time Domains are the subject of remote viewing. The Memory domains are the subject of psychology. This book is only discussing the scientific theories of remote viewing.

Mind-browser Windows And Attention

The following is what my mind-browser was open to earlier this morning.

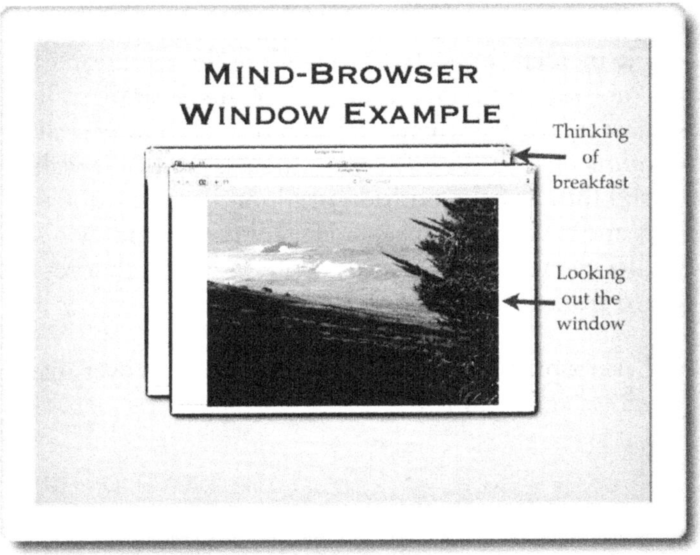

Here is what I saw in my mind-browser. I was standing on my balcony looking at the ocean. I was in a meditative state so my mind was focusing on what I was viewing. I did have a little attention in a second window because I was also thinking about getting some breakfast.

I know a man who is a Vietnam war veteran who has a foreground window for what he is doing now. However, he has Post Traumatic Stress Syndrome and most of his attention is on a "Vietnam in 1967" window. He is momentarily expecting machine gunfire or a mortar attack and is ready to take cover. If he hears any unexpected loud noise, he switches to the Vietnam window and is ready for hand-to-hand combat. His attention is always ninety-percent on Vietnam.

Recently, I was talking to a man who related to me the story of a life-changing motorcycle accident he had forty years ago. He told me how he could replay the whole incident in his head, including the sensation of the motorcycle going out of control, his position as he slid across the pavement, the sound of the motorcycle grinding against the pavement, the texture of the wall he slid into and the position of his body as he came to rests and lost consciousness. He still has a window open to that event and lives with the physical pain that resulted from the accident.

Other people have their attention spread out over many windows at the same time.

In this example, the person might be talking to the woman in the foreground window while they share their attention with other places and times. These other times or places are not random events. They are all of interest for some personal reason.

The process of remote viewing requires people to get into a quiet state, close all of their browser windows, and concentrate on one window and a single ERL. This requires a meditative state.

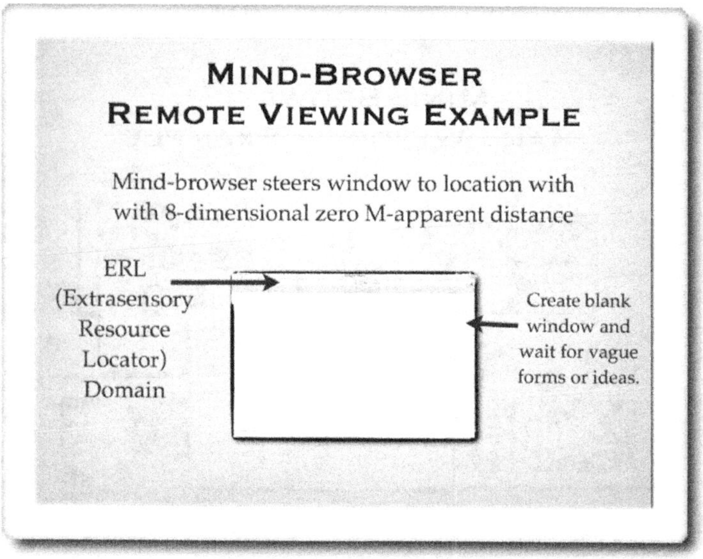

To do remote viewing, one must start by creating a blank mind-browser window and quieting their thoughts. Then, they turn their attention to an ERL, such as those described in the following section, a target. Their mind-browser steers the window to a location with zero M-apparent distance in 8-dimensional Minkowsky space. Then, they wait for vague forms and ideas to appear in their browser window.

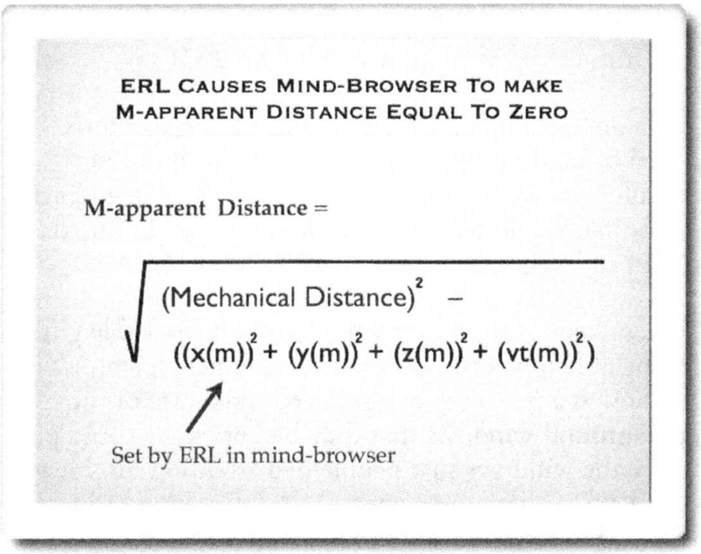

I have no reductionist scientific explanation for how our mind does this. It is an emergent phenomenon that has been extensively demonstrated as described in the preceding section. It is also common practice for everyone to do this in a limited sense. For instance, if I ask you to visualize what you had for lunch yesterday, your mind gets the *x,y,*and *z* coordinates for the restaurant and the *t* coordinate for noon yesterday. Your mind-browser goes to the zero M-apparent distance to your lunch.

In the browser-mind metaphor, long term memory includes only “browser windows” that were of special interest and were bookmarked. These “window bookmarks” may be special events such as a wedding day, traumatic events such as breaking a leg skiing, emotional events such as the time your first love dumped you or, watching a disaster or accident. You probably don’t have a bookmark for what you were doing on

September 8th in 2001, but you probably have one for the images and your activity on 9/11/2001.

One does not have to worry about having windows open to random places or events. One will not randomly experience being on the sinking Titanic or at the assassination of Archduke Franz Ferdinand of Austria in 1914 unless one chooses that ERL.

I once knew a counselor who had the remarkable gift of being able to see the pictures in other people's browsers. Although a person may not have been aware of the background windows that they had open, he could perceive the windows that people had attention on. He was excellent at directing a person's attention to past incidents that were causing problems in the now.

Example ERLs

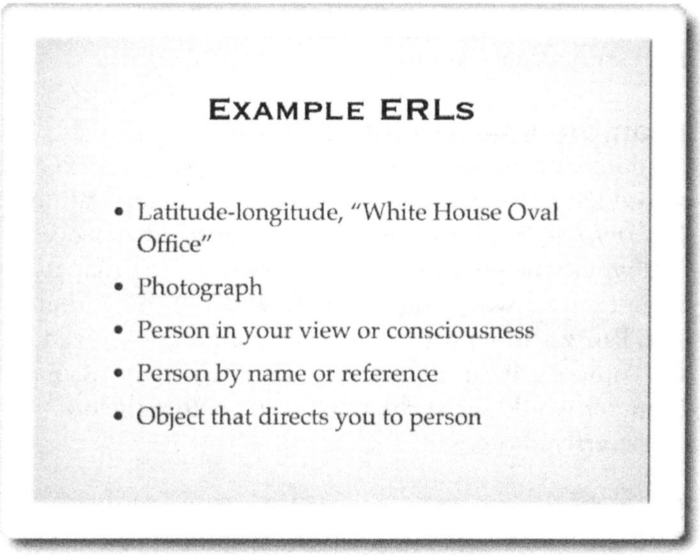

In a remote viewing, there can be many kinds of ERLs. Intelligence remote viewing targets could be simply a latitude or longitude in numbers like 38 degrees North, 125 degrees West. The skilled remote viewer's mind-browser could somehow be directed to that location and they could describe the vague images and perceptions of that location. A remote viewer's mind-browser could be directed to a specific place such as the White House Oval Office. These types of ERLs generally require people of extraordinary natural ability or extensive training.

Other people could be shown a photograph of something and be able to explain where it is or something about it. I met a healer, Jixing Li, who, in Seattle, was shown a photograph of an incubator for culturing cancer cells located in a laboratory in Pennsylvania. He could discriminate between cancer cell cultures in the top and the bottom of the incubator, and alter the growth[97] of one group and not another. He, of course, has extraordinary ability.

Example Space-Time ERLs

Remote viewing of future or past events requires extraordinary ability or extensive remote viewing training. In *Limitless Mind,* Russell Targ related the experience at SRI where the gifted viewer, Pat Price, was participating in an exercise where another man was sent to a remote site. Pat was to report what the outbounder observed. Pat could easily time shift so he reported what the outbounder would see at the remote site before the observer arrived there.

In another experiment, Russell Targ with others successfully forecast the silver futures market. Their story made the front page of Wall Street Journal. A businessman wanted to know whether the silver futures market would go "up-a-little," "up-a-lot," "down-a-little," or "down-a-lot" a week ahead of time, let's call it June 15. Without the remote viewers knowledge, on June 7 the businessman selected four dissimilar objects and associated each of them with each of the four market outcomes. Depending how the market went, one of the four would be shown to the remote viewer on June 15 at ten in the morning at Joe's Coffee House. If the market went "up-a-lot," he would show the remote viewer a Teddy Bear. If the market went "down a lot," he would show him a light bulb, etc.

Then, on June 8, without knowledge of what the businessman's four objects were, the remote viewer was asked to describe what he would be shown on June 15 when he met with the businessman at Joe's Coffee House at ten in the morning.

The remote viewer would set his ERL for ten in the morning on June 15 and Joe's Coffee House. (This is the x,y,,z, and *t*.) If, on June 8, the remote viewer described "something brown and fuzzy that had a squeaker inside" the businessman would know the market was going "up a lot" and make a buy-long order. Then, on June 15, if the market went "up a lot" he would show the remote viewer the Teddy Bear.

If the remote viewer described "something round, white, and smooth that was partly metal" on June 8, the businessman would know the market was going "down a lot" and make a short-sale order. Then, on June 15, if the market went "down a lot" he would show the remote viewer the light bulb.

In nine straight weeks the remote viewing was correct, and the group made $100,000.

I suspect that many are repeating a refined version of the experiment today and not talking or writing papers about it.

In the Pat Price example, the ERL was the person being sent out to do the remote viewing at a designated *t*. Pat simply put that ERL in his browser and reported what was in the "window."

In the silver commodity trading example, the remote viewer set his ERL to one week later (*t*) at the place (*x,y,z*) where he agreed to meet the businessman.

Somehow, they tuned their "browsers" so that they perceived events in the future.

Time As An Illusion

> ***"...for us physicists believe the separation between past, present, and future is only an illusion, although a convincing one."***
>
> ***Albert Einstein[98]***

> ***Time-space is a convenient way of not having everything happen all at once in the same place.***
>
> ***Anonymous***

Despite the above quote that time as an illusion, it is still necessary to go beyond Einstein's Horizon in thinking about time. When Einstein published his paper on Special Relativity in 1905, he revolutionized physicist's thinking about time and undid the ideas of time as an absolute that had existed since Newton. Einstein introduced the idea that your concept of time depends on what you are observing and how fast you are going.

In remote viewing, time is not relative in Einstein's sense because the remote viewer is not traveling near the speed of light. However, your mind-browser can make the M-apparent distance equal to zero for any present or future time. Both past and future are accessible to practiced or naturally gifted viewers.

In both Physics and Metaphysics, time is considered an illusion. Roger Penrose's book[99] devotes two lengthy, mathematical chapters on *Spacetime* and *Minkowskian*

Geometry. Here, I will give only the general conceptual idea.

On the LCD monitor of my computer, there is a small black spot in the center of the screen where a screen pixel has failed. That spot appears in the center of my browser window. If I stream a movie about the sinking of the Titanic, the whole movie passes and the failed pixel doesn't get wet. The movie is information. The failed pixel is physical reality. Information is separate from physical reality. In the 194 minute movie, the story spans about fifty years. While I streamed the movie, the failed pixel aged 194 minutes. Physical reality chugged along while, in the information reality, everybody drowned.

The 8-dimensional Minkowsky space-time reality is like a big online video store. The movie (information) of everything that ever happened and will happen exists there online. I can experience the (information) sinking of the Titanic by streaming the movie. However, while I watch the movie, the physical reality Titanic is a rusting hulk at the bottom of the ocean.

Information exists in 8-dimensional complex Minkowsky space. Physical reality exists in only four dimensional spacetime.

The separation precludes physical time travel. You can access the information of the past or future through remote viewing. You can't become physically a part of or physically interact with those past or future frames by remote viewing.

Remote viewing does not deal with physical time travel or any energy transfer in space-time. Speculation about physical time travel often deals with conundrums like, "if I time traveled back and shot my father before I was conceived, I would not exist. Therefore, I couldn't have existed to go back and shoot him." In remote viewing you only can only go back in time and view your father and possibly convey information about why he should not marry your mother. But, he still will.

In the remote viewing example of predicting the silver futures market, the businessman agreed to show the remote viewer an object (e.g. Teddy Bear) a week in the future which depended on whether the market went up or down. What the remote viewer saw was information. The knowledge of the future market was either used or not used by the businessman. The market did what it did independently of the businessman's knowledge.

A man has scheduled a flight for a business trip. He has a dream of being on an airplane that crashes and burns. He uses that information to cancel his flight. He subsequently finds out that the flight he was scheduled to be on did indeed crash, burn, and everyone on board was killed. The airplane did what it did, independently of the man's knowledge.

In both examples the future happened as predicted by remote viewing. However, the people used their knowledge to make a decision to participate or not participate in that future.

Some schools of thought believe that remote viewers can only obtain information from minds making or having made an observation. When the remote viewer saw the Teddy Bear, his mind-browser was synchronizing with his own one-week-future mind-browser which saw the Teddy Bear. In the case of the plane crash, the remote viewer's dreaming synchronized his mind-browser with someone on the airplane experiencing the crash. It wasn't his mind-browser because he didn't take the flight.

In another example, someone could receive a large unmarked envelope containing a Hawaiian travel brochure. Since they did not know who sent the envelope, they couldn't determine what is in the envelope by focusing on the source. However, they could remote view the contents by focusing on their future self, two minutes later, examining the contents of the envelope. Their mind-browser might come up with a "lush, green, blue, happy," prediction of what is in the envelope.

I live in a rural area where our mailboxes are on posts at the side of the street. After the mailman passes, I could wish to know what is in my mail. Since I am not a gifted psychic or highly trained remote viewer, I cannot look down the driveway and use "X-ray vision" to see what is in the mailbox. I can go into a meditative state and then, in my mind-browser, imagine myself standing in front of the mailbox in a few minutes, holding the mail. I might perceive and draw something rectangular and thick like a box. I might view myself shaking the box and I could write down that something rattled in the box. I might also draw another rectangular object, maybe something like a pile of catalogs and magazines. I could continue this for quite a while and gain a lot of information. However, it's easier to simply walk to the mailbox and look.

In *Memoirs of A Psychic Spy*, Joe McMoneagle relates a story about doing a live TV remote viewing demonstration. He was to remote view the contents of one of four envelopes. Each envelope contained a photograph of a landmark in the city. The envelope he was to view was to be selected during the filming of the show. Because he was concerned about the distractions of the studio during the show, he did the remote viewing early in the day of the show, before the envelope was selected. He was successful in telling the TV interviewer the correct contents. He was able to do the correct viewing in the morning, even though the envelope had not yet been selected.

You don't have to be traveling near Einstein's relativistic speed of light to do any of this. Remote viewing in space-time can be done sitting in your chair.

Vagueness And Noise in Remote Viewing

Most remote viewers get a rather vague impression of their targets. A few gifted practitioners who are generally identified as psychics can produce clear, unambiguous descriptions of what they are viewing.

In communication systems, performance is usually described in terms of bandwidth, signal, and noise. Systems with a lot of noise relative to the signal generally have higher error rates or less information. In hardware communication systems the noise is usually electromagnetic "static" or the thermal noise added by the equipment. The bandwidth limits the rate of transmission of information. DSL and cable modems have broadband circuits and operate at megabits per second. Older Wi-Fi and cell phone text systems have lower bandwidths and are slower.

If the remote viewing mechanism is synchronization, as described in Section IV, the frequencies are those of the brain nervous activity. These are in the delta, theta, alpha, or beta range: 0-24 Hz. In this frequency range, the bandwidth is very low and the data rates are small. The information must be vague and require a while to acquire.

In Remote Viewing The "Noise" Comes From Other Sources.

Judgement or Assumption Noise

Here, I paraphrase a remote viewing session from Courtney Brown's book.

If I gave a remote viewer the address of 625 Nelson Road, Stanford, CA, on a fall weekend afternoon, their report might be'

"A large rectangle, a building–surrounded by trees–no roof to the building–like a big courtyard.

Many people in the courtyard–high emotions–some anger and frustration. A riot or demonstration–charismatic leaders–cheering. Things waving."

The following is the satellite view of the example site, however, it is not a picture taken on a Saturday afternoon during football season.

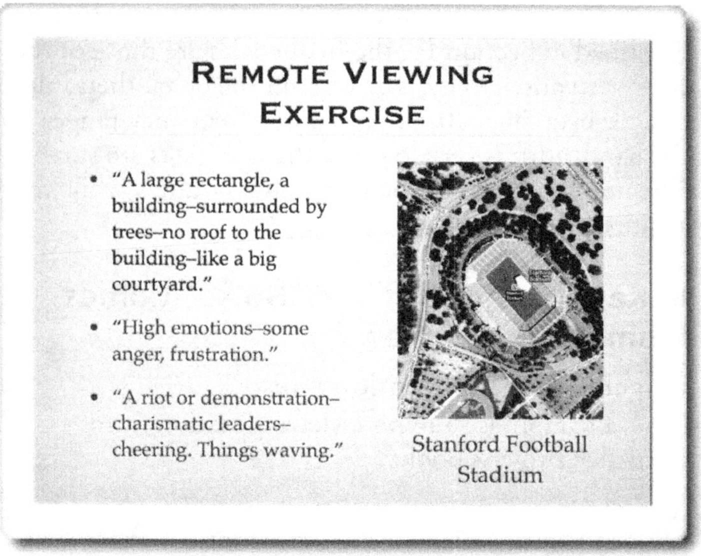

The remote viewer would be focused on the whole target. The football game itself would be subsumed into the crowd activity. It would be difficult for the viewer to zoom in closer. The activity would become interference.

The viewer's description of the target displays the discipline of not making an evaluation of "what it must be." Adding an analytic overlay of judgement or assumptions would add "noise" to the process. For instance, he could have added the judgement noise that it must be a bullfight, a political rally, or a baseball game. Any such judgement or evaluation would have degraded the report–added "noise."

Space-Time Focus Noise

Remote viewing requires an ERL to be given to the mind-browser. If it gets a precise *x,y,z*, and a very vague *t*, noise will be added. For instance, if I gave a viewer the *x,y,z,* for a downtown Manhattan location and the whole year of 2001 they might sense general chaos and movement. It would be very vague and details would be noisy. If I limited *t* to September they might vaguely sense grief and disorientation "colored" noise. If I limited the *t* to the morning of 9/11/2001 they would sense clear disbelief and then clear terror etc.

Similarly, too large a focus of space can add noise. In the preceding example of 9/11, if the *x,y,z,* was the entire state of New York, the feeling might have been that of noisy confusion.

In the previous example of the remote viewer and the silver futures businessman, the task was to remote view a specific future time and place. The revelation of the Teddy Bear or light bulb was to take place at ten AM at Joe's Coffee House at 680 Main Street. That specification reduces the noise or ambiguity.

If the remote viewer were not given a specific time to view the football field, (during the game) they might have observed it during a time when only the groundskeepers were in the stadium. The result would have been quite different.

---

97 Neely, John. "Alteration in Cultured Cancer Cell Growth through Trans-Continental Healing Intention and Sustained Energy Imprinting." *ISSSEEM 18th Annual Conference: Energy, Intent, and Healing, Boulder* CO, June 2008 Print

98 A condolence letter to the family of his departed friend Besso. He said that although Besso died first, it was of no consequence.

99 Penrose, *Road to Reality*. See #27, above.

VII Dealing With The Bigot's Protocol

Now, you know a lot about remote viewing and have some confidence that it has some scientific basis and is certainly not "pseudoscience and superstition," as some critics have said. I have "tried the case" for and against the validity of remote viewing and I trust your verdict is in favor of the validity. I also trust that your vision extends beyond Einstein's Horizon.

If you happen to mention remote viewing in the presence of someone who thinks they know everything about science, they might start to run the Bigot's Protocol on you. Now, it can go like this:

You mention that you have recently read a book about remote viewing.

After the Bigot's face begins to flush and their jaw tighten, and as they start raising their finger to shake at you, say:

> "Isn't it exciting that, after some seventy years, physicists have solved the Einstein–Podolsky–Rosen paradox about entanglement in Quantum Physics! Who would think that the 8-dimensional Minkowsky geometry would also explain why remote viewing works? I'll be happy to email you the Rauscher-Amoroso and Rauscher-Targ papers on that.
>
> It is really hard to not let one's thinking become obsolete these days. If one doesn't keep up, it is easy to end up thinking your world is flat."

Download the papers [100] and email them. And,

don't forget to tell them where to buy this book.

---

[100] Available with a Google search of: "Relativistic Physics in Complex Minkowski Space, Nonlocality, Either, Model and Quantum Physics" And search: "The speed of thought: Investigations of a complex space-time metric to describe psychic phenomena"

About The Author

I was Chief Scientist and Marketing Manager at a Billion dollar company that manufactured communications satellites. We launched them on the Space Shuttle and other rockets. My job was to sell commercial satellites to companies that used them to broadcast TV and telephone services. AT&T and its foreign counterparts were typical customers.

The trouble with selling satellites is nobody can see them. I couldn't take a sample case around or sell them out of the back of my car.

All I had to sell were the beliefs about satellites we could build. At the end of our four-year sales campaign, I wanted my customers to hold a large number of positive beliefs about our product: our company had more successful satellites in space than our competitors, our management was better than the competitors', customers could make a lot of money from our product, and so forth. I had to orchestrate the communication of beliefs about good design from about a hundred of our technical specialists (most with technical PhDs) with their counterparts in the customers' organizations.

Their beliefs were of the best way to build solar cells, transmitters, batteries, etc. I also had to sell the beliefs of who the specialists were as personalities: that our guys would be good to work with in overseeing the four-year design and building of the satellite.

I had to understand large scale systems and be able to grasp the technical essence of many specialties. Working in a hight tech firm allowed me to continually be exposed to new ideas and learn to understand the essences of theoretical arguments. Many times, I would spend days working with a specialist in a technical area distilling the key ideas of their technical achievements or research priorities. If a customer wanted to know about the latest research in battery electrolytes, we would create a presentation on the subject. If a customer wanted to know about the latest radio modulation technology, we would create a pitch

My own technical specialty is communication theory. I hold patents on satellite antenna systems. I hold a patent on a satellite (music) broadcast receiver for automobiles. I hold a patent for a circuit that may be in your cell phone. I managed the development of the first CDMA mobile telephone.

I used these diverse skills in preparing *Science, Remote Viewing, and ESP: Beyond Einstein's Horizon.*

I have a BSEE degree from Stanford.

Appendix On Holograms

The following diagrams further illustrate how the same information is not available in all parts of a hologram.

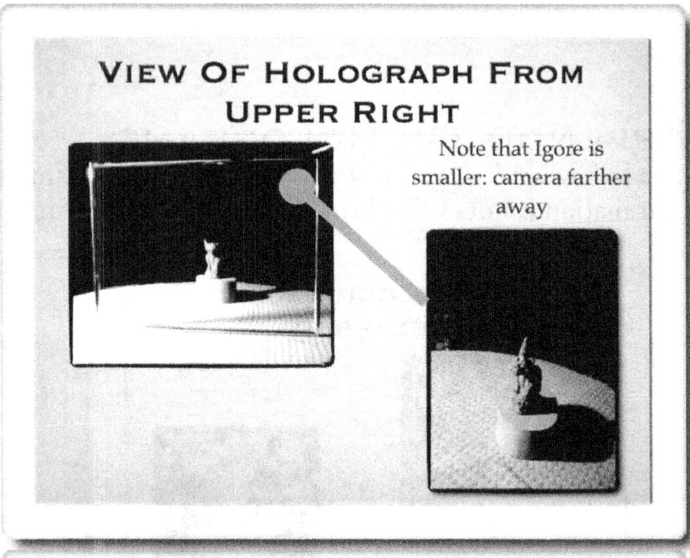
VIEW OF HOLOGRAPH FROM UPPER RIGHT
Note that Igore is smaller: camera farther away

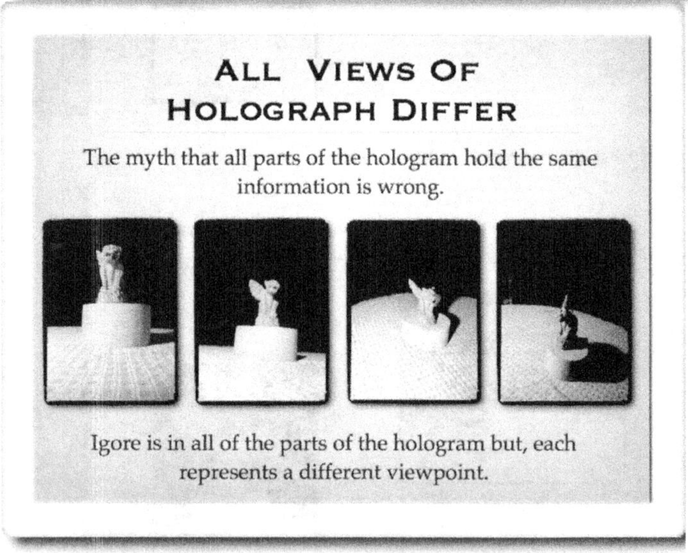
ALL VIEWS OF HOLOGRAPH DIFFER
The myth that all parts of the hologram hold the same information is wrong.
Igore is in all of the parts of the hologram but, each represents a different viewpoint.

BIBLIOGRAPHY

Atwater, F. Holmes. *Captain of My Ship, Master of My Soul.: Living With Guidance.* Charlottesville: Hampton Roads , 2001. Print.

Backster, Cleve, and Stephen White. "Biocommunication Capability: Human Donors and in vitro leukocytes." *International Journal of Biosocial Research* 7(1985): 132-46. Print

_______. *Primary Perception,*. Anza: White Rose Millennium P., 2003. Print

_______. "Plants Can Sense Human Intention." YEN, YAP WAI . Spiritual Village, 08 May 2008. Video. 25 Feb 2009 <http://spiritualvillage.blogspot.com/2008/05/video-cleve-backster-vs-mythbuster.

Bell,J.S. "On the Einstein-Poldolsky-Rosen Paradox", *Physics* 1 (1964):195 (1964). Print

Blavatsky, H. P. *The Secret Doctrine: The Synthesis of Science Religion and Theosophy*, London, 1888. Print

Braud, William. *Distant Mental Influence: Its Contributions to Science, Healing, and Human Interactions.* Charlottesville: Hampton Roads Pub., 2003. Print

Brown, Courtney. "Differential Equations: A Modeling Approach." *Quantitative Applications in the Social Sciences series* 150: Sage Pub, Thousand Oaks. (2007) Print

_______. "Graph Algebra: Mathematical Modeling with a Systems Approach" *Quantitative Applications in the Social Sciences series* 151: Sage Pub, Thousand Oaks. (2008) Print

_______. *Ballots of Tumult: A Portrait of Volatility in American Voting.* Ann Arbor: U of Michigan P., 1991. Print

_______. "Chaos And Catastrophe Theories." *Quantitative Applications in the Social Sciences series*, 107. Thousand Oaks: Sage. 1995. Print

_______. Remote Viewing: *The Science and Theory of Nonphysical Perception.* Atlanta: Farsight P., 2006. Print

_______. *Serpents in the Sand: Essays on the Nonlinear Nature of Politics and Human Destiny.* Ann Arbor: U. of Michigan P., 1995. Print

Cohen, D, et al. "New Six-Layer Magnetically-Shielded Room for MEG." Biomagnetic Proceedings.. (Massachusetts General Hospital) 01Jan2002. Print

Dubrov, A.P., and V.N. Pushkin. Parapsychology and Contemporary Science. New York and London: Consultants Bureau, 1982. Print

Einstein, A., B. Podolsky, and N. Rosen, "Can a quantum mechanical description of physical reality be considered complete?" Physical Review, 47, (1935) 777-80. Print

_______, and Marcel Grossmann,. Entwurf einer verallgemeinerten Relativitätstheorie und einer Theorie der Gravitation Leipzig: Teubner, 1913. Print

_______. "Zur Elektrodynamik bewegter Körper," Annalen der Physik 17 (1905): 891-21. Print

Everett, Hugh, III. "'Relative state' formulation of quantum mechanics." Review of Modern Physics 29:(1957): 454–62. Print.

Gabriel, S., and. Gabriel, R. , C "The Dielectric Properties of Biological Tissues II: Measurements in the Frequency Range 10 Hz – 20 GHz," Physics of. Medical Biology, 41: 2251-69,(1996) Print

Hartley, R. V. L., "Transmission of Information". Bell System Technical Journal. (July 1928) Print

Hunt, Valerie. Infinite Mind: Science of the Human Vibrations of Consciousnesses. Malibu: Malibu Pub, 1989. Print

Hyman, Ray. "Evaluation of a Program on Anomalous Mental Phenomena". Journal of Society for Scientific Exploration, 10.1: Article 2. Print

ISSSEEM 18th Annual Conference: "Energy, Intent, and Healing.".Boulder CO, June 2008

Johnson, D.D. "Twins Separated at Birth: The Story of Jim Lewis and Jim Springer." Smithsonian Magazine Oct 1980. Print

Klebanoff, Nina A., and Patsy K. Keyser. "Menstrual Synchronization." Journal of Holistic Nursing, 14. 2,(1996): 98-114. Print.

Kut, Carmen, Vahid Golkhou, and Joel S Bader. "Analytical approximations for the amplitude and period of a relaxation oscillator." BMC Systems Biology, 3:6 (2009) doi:10.1186/1752-0509-3-6 (2009). Print

Laughlin, Robert. "FRACTIONAL QUANTIZATION." Dep. of Physics, Stanford University Dec. 8 1988 Nobel Lecture, Print

_______. A Different Universe: Reinventing Physics From the Bottom Down. New York: Basic Books, 2005. Print

Maurer, Leon H. "Einstein And The Secret Doctrine." ABC Images, Oct 27, 1997.Web. 13 Apr 2009 <http://leonmaurer.info/ABCimages/einstein.html>.

May, Edwin C. . "The American Institutes for Research Review of the Department of Defense's STAR GATE Program: A Commentary." *The Journal of Parapsychology.* 60(March 1996): 3-23. Print

McCraty R. "Influence of cardiac afferent input on heartbrain synchronization and cognitive performance" Int. *Journal Psychophysiology*; 45 (2002):72-73, Print

_______. and Atkinson M. "Psychophysiological coherence" *Emotional Sovereignty.* Eds. Childre D. and McCraty R, Wilson BC, Amsterdam: Harwood Academic Pub., forthcoming.

_______ , Mike Atkinson, Dana Tomasino, and William A. Tiller. "The Electricity of Touch: Detection and Measurement of Cardiac Energy Exchange Between People ." *Brain and Values: Is a Biological Science of Values Possible.* Ed.Karl H. Pribram, Mahwah: Lawrence Erlbaum, 1998. 359-79. Print.

_______. *The Energetic Heart: Bioelectromagnetic Interactions Within And Between People. Boulder Creek,*: Institute for HeartMath, 2003 Print.

McMoneagle, Joseph. *Memoirs of a Psychic Spy:.* Charlottsville: Hampton Roads Pub, 2006. Print

Miklavic D., N. Pavs, and F. X. Hart, "Electric Properties of Tissues," *Wiley Encyclopedia of Biomedical Engineering*, Hoboken, Wiley, 2006. Print

Minkowski, Hermann. "Das Relativitätsprinzip." Undated T. S. . Cod. Ms. Math. Arch. 60: 3. Göttingen:. Niedersächsische Staats und Universitätsbibliothek, (1907) Print

Pelting et. al. "Local Nanomechnical Motion of the Cell Wall Saccharomyces cervisiiae" *Science* 305, (20 Aug. 2004) Print

Penrose, Roger. *The Road to Reality: A Complete Guide to the Laws of the Universe.* New York: Knoff, 2005. Print

Pikovsky, Arkady, Michael Rosenblum, and Jurgen Kurths. *Synchronization: A Universal Concept in Nonlinear Science.* Cambridge: Cambridge, (2003) Print

Puthoff, H.E, and Russell Targ. "Information Transmission under Conditions of Sensory Shielding." *Nature* 251 (1974): 602-07. Print.

_______, and R. Targ. "A perceptual channel for information transfer over kilometer distances: Historical perspective and recent research.." *Proceedings of the IEEE.* 64.3(March 1976): 329- 54. P

_______. "CIA-Initiated Remote Viewing Program at Stanford Research Institute." *Journal of Scientific Exploration.* 12,1(Spring 1996):Article 3. Print.

Ramachandran, Vilayanur . "40/40 Vision Lecture: Neurology and the Passion for Art." YouTube. 04/20/2009). U of California at San Diego. 21 Apr 2009 Web <http://www.youtube.com/watch?v=0NzShMiqKgQ&fea

_______, Interview "The Human Face", (2001). DVD, ISBN 0-7907-6038-X.

Rauscher, E.A. Electromagnetic Phenomena In Complex Geometries and Nonlinear Phenomena, NonHertzian Waves and Magnetic Monopoles, Milbrae:Tesla, 1983, Print

_______. "Non-Abelian gauge groups for real and complex amended Maxwell's equations." Eds. R.L. Amoroso, G. Hunter, M. Kafatos, and J-P Vigier. Gravitation and Cosmology: From the Hubble Radius to the Planck Scale, Dordrecht: Kluwer Academic, 2002, 183-188; Print

_______. and R. Targ, "Investigation of a complex space-time metric to describe precognition of the future," Frontiers of Time: Retrocausation – Experiment and Theory, ed D.P. Sheehan. Melville: AIP Conf. Proc.. (2006) Print

_______. and R.L. Amoroso, "Relativistic Physics in Complex Minkowski Space, Nonlocality, Either, Model and Quantum Physics" Physical Interpretations of Relativity Theory: London, Imperial College, British Society for the Philosophy of Science Conference, 12 Sept. 2008. Print

_______. and Targ, R. "The speed of thought: Investigations of a complex space-time metric to describe psychic phenomena", Journal of Scientific. Exploration, 15,(2001) 331. Print

_______. Electromagnetic Phenomena In Complex Geometries and Nonlinear Phenomena, NonHertzian Waves and Magnetic Monopoles, Milbrae:Tesla, 1983, Print

_______, R.L. Amoroso, “The Physical Implications of Multidimensional Geometries and Measurement”, International J of Computing Anticipatory Systems, Liege, Belgium. CHAOS,pub. 2006 1, Ed. Daniel M. Dubois, Published by ISSN 1373-541. Print

_______. *Electromagnetic Phenomena In Complex Geometries. Oakland.* Noetic P., 2008. Print

Rawson, Rosemary . "Two Ohio Strangers Find They're Twins at 39—and a Dream to Psychologists." PEOPLE magazine, May, 1979. Print

Remote Viewing History, Facts, Names, Places and Events." Psi Tech News. Psi Tech Inc. 12 Apr 2009. Web <http://www.psitech.net/news/therealremoteviewinghistory.htm>

Rosemblum, Michael G., Arkady S. Pikloski, and J Urgen Kurths. "Synchronization Approach To Analysis Of Biological Systems" *Fluctuation and Noise Letters*: FLN4.2,Print (2004), World Scientific Pub.(2) Print

Salart, D.; A Baas, J. A. W. van Houwelingen, N. Gisin, and H. Zbinden, “Spacelike Separation in a Bell Test Assuming Gravitationally Induced Collapses.” *Physical Review Letter* 100, 220404 (2008).Print

Schulz, Mona Lisa. *Awakening Intuition: Using Your Mind-Body Network for Insight and Healing.* New York:Harmony, 1998. Print

_______. *The Intuitive Advisor: A Psychic Doctor Teaches You How to Solve Your Most Pressing Health Problems.* Hay House, 2009. Print

Shannon, C. E. “A Mathematical Theory of Communication”, *Bell System Technical. Journal.*, 27 (1948) 379-423, 623-56, Print

Strogatz, Steven H. *Sync: The Emerging Science Of Spontaneous Order.* New York: Hyperion, 2004. Print.

Targ , R., and H.E. Puthoff. *Mind-Reach: Scientists Look at Psychic Abilities (Studies in Consciousness).* NewYork. : Delacorte, 1977. Print.

_______. "Remote Viewing at Stanford Research Institute in the 1970s: A Memoir." *Journal of Scientific Exploration* 10,1(1996): 77-88. Print.

_______. *Limitless Mind: A Guide to Remote Viewing. And Transformation of Consciousness* Novato: New World Library, 2004. Print

Tart, Charles T., Harold E. Puthoff, and Russell Targ, eds. *Mind At Large:Institute of Electrical and Electronic Engineers Symposia On The Nature of Extrasensory perception.* New York, Praeger, 1979. Print

Tompkins, Peter, and Christopher, Bird. *The Secret Life of Plants.* New York.: Harper and Row, 1973. Print

Warcollier, Rene. *Mind to Mind.* Charlottesville: Hampton Roads Pub., 1948. Print

Whitson, T.W., D.N. Bogart, J. Palmer, and C.T. Tart. "Preliminary experiments in group Remote viewing." *Proceedings of the IEEE.* 64.10(Oct. 1976): 1550- 51. Print.

Wiener, Norbert , *Cybernetics or Control and Communication in the Animal and the Machine,* Paris: Hermann et Cie - MIT Press,(1948) Print

Willard B. Gatewood, Jr, *Controversy in the Twenties; Fundamentalism, Modernism, and Evolution.* Nashville: Vanderbilt U. P, 1969. ix + 459. Print

INDEX

D

E

F

G

H

Made in the USA
Monee, IL
24 November 2025

36247473R00144